Military Space Power

**Recent Titles in
Contemporary Military, Strategic, and Security Issues**

Information Operations—Doctrine and Practice: A Reference Handbook
Christopher Paul

The National Guard and Reserve: A Reference Handbook
Michael D. Doubler

Returning Wars' Wounded, Injured, and Ill: A Reference Handbook
Nathan D. Ainspan and Walter E. Penk, editors

Manning the Future Legions of the United States: Finding and Developing Tomorrow's Centurions
Donald Vandergriff

The Process and Politics of Defense Acquisition: A Reference Handbook
David S. Sorenson

International Crime and Punishment: A Guide to the Issues
James Larry Taulbee

Serving America's Veterans: A Reference Handbook
Lawrence J. Korb, Sean E. Duggan, Peter M. Juul, and Max A. Bergmann

Military Doctrine: A Reference Handbook
Bert Chapman

Energy Security Challenges for the 21st Century: A Reference Handbook
Gal Luft and Anne Korin, editors

An Introduction to Military Ethics: A Reference Handbook
Bill Rhodes

War and Children: A Reference Handbook
Kendra E. Dupuy and Krijn Peters

Military Justice: A Guide to the Issues
Lawrence J. Morris

Military Space Power

A Guide to the Issues

Wilson W. S. Wong
and
James Fergusson

Contemporary Military, Strategic, and Security Issues

PRAEGER

AN IMPRINT OF ABC-CLIO, LLC
Santa Barbara, California • Denver, Colorado • Oxford, England

Copyright 2010 by Wilson W. S. Wong and James Fergusson

All rights reserved. No part of this publication may be reproduced, stored in a retrieval system, or transmitted, in any form or by any means, electronic, mechanical, photocopying, recording, or otherwise, except for the inclusion of brief quotations in a review, without prior permission in writing from the publisher.

Library of Congress Cataloging-in-Publication Data
Wong, Wilson.
  Military space power : a guide to the issues / Wilson W. S. Wong and James Fergusson.
    p. cm. — (Contemporary military, strategic, and security issues)
  Includes bibliographical references and index.
    ISBN 978-0-313-35680-3 (alk. paper) — ISBN 978-0-313-35681-0 (ebook)
 1. Astronautics, Military—Handbooks, manuals, etc.   2. Space weapons—Handbooks, manuals, etc.   3. Space warfare—Handbooks, manuals, etc.
    I. Fergusson, James G. (James Gordon), 1954–   II. Title.
    UG1530.W646 2010
    358'.8—dc22   2009051219

ISBN: 978-0-313-35680-3
EISBN: 978-0-313-35681-0

14 13 12 11 10   1 2 3 4 5

This book is also available on the World Wide Web as an eBook.
Visit www.abc-clio.com for details.

Praeger
An Imprint of ABC-CLIO, LLC

ABC-CLIO, LLC
130 Cremona Drive, P.O. Box 1911
Santa Barbara, California 93116-1911

This book is printed on acid-free paper ∞

Manufactured in the United States of America

Contents

Introduction	The Evolution of Military Space	1
Chapter 1	The Scientific and Technological Foundations of Military Space	14
Chapter 2	Military Space and Force Enhancement	41
Chapter 3	Military Space and Force Application I: Space Surveillance and Passive Measures	66
Chapter 4	Military Space and Force Application II: Active Force Application	90
Appendix	Treaty on Principles Governing the Activities of States in the Exploration and Use of Outer Space, Including the Moon and Other Celestial Bodies	117
Glossary: Military-Related Terms, Activities, Events and Technologies		123
Bibliography		145
Index		153

INTRODUCTION

The Evolution of Military Space

Outer space, or the heavens has long captured the imagination of humankind. Understood for millennia in religious or theological terms, around the beginning of the 20th century, writers of science fiction, a relatively new genre of literature, began to speculate on the future exploration and exploitation of outer space. Within this new genre as it evolved over the century, writers transposed their ideas of relations among societies, including war, into the outer-space domain. As science fiction moved into the new media of film and then television, these ideas about the future, including military clashes in outer space, were popularized among a wider audience, especially in North America. Today, much of the public understands space through the lens of popular television series such as *Star Trek* and films such as *Star Wars*. Indeed, in an interesting twist, Ronald Reagan's famous Strategic Defense Initiative (SDI) became known as Star Wars in an attempt to discredit it as pure science fiction.

Speculation slowly began to turn into reality with the development of rocket technology—the prerequisite for exploring and exploiting outer space—and two broad schools of thought emerged. The first saw space largely in scientific and explorational terms, perhaps most clearly encapsulated with the creation of the National Aeronautical and Space Administration (NASA) during the Eisenhower administration and John F. Kennedy's New Frontier. The hope of this school of thought was that space exploration would be driven by scientific curiosity for the benefit of all humankind. The conflicts among societies and states would be put aside in favor of global cooperation for the common good. Outer space would remain a pristine environment or domain, unpolluted by weapons and war.

The other school of thought saw the future of outer-space exploration and exploitation in terms of the natural rivalry between societies and states common to history. Space could not be divorced from the political reality of the world. While regrettable for many, the idea of outer space as the ultimate high ground promised a distinct advantage to states able to exploit it for terrestrial purposes, as reiterated in the most recent *U.S. National Space Policy*: "those who effectively utilize space will enjoy added prosperity and security and will hold a substantial advantage over those who do not."[1] This primarily meant the great powers, or the superpowers of the post–World War II era, which possessed the resources to develop the means to do so. As no power could allow its rival such a distinct advantage, competition and conflict were inevitable. The cold war rivalry between the United States and Soviet Union would logically extend into space. The race to place the first satellite in orbit (1957) and the first man into space (1961), both won by the Soviet Union, and to visit the moon (1969), won by the United States, was more about this rivalry than simple scientific curiosity.

While rocketry dates back to antiquity in the form of Chinese gunpowder rockets, modern rocket science can be traced back to Sir Isaac Newton's three laws of physical motion; Colonel William Congreve's development of the Congreve rocket, referenced in William Scott Key's line "the rocket's red glare" in "The Star-Spangled Banner"; and the pioneering work of Konstantin Tsiolkovsky (Russia), Robert Goddard (the United States), and Hermann Oberth (Germany). It was national security and military considerations that provided the rationale for devoting large amounts of national resources to develop a modern rocket. The first operational liquid-fueled, long-range rocket for military use was the German V-2 armed with a conventional warhead and launched at targets in Western Europe in early September 1944. Since then, national security and military considerations, alongside related notions of national prestige, have been the primary drivers behind the development of sophisticated rockets and satellites performing three major functions—surveillance or earth observation, global communications, and global navigation. Both scientific or civil and commercial use followed from, and were a product of, initial military-security investments. The story of outer space from its origins to today, and into the future, is largely the story of military use.

Understanding Military Space

The tendency to conceptualize space into the categories of peaceful and military use is long-standing, yet problematic. Both employ the same basic physics and technology relative to the four basic components of space use: the rockets that provide access to space; the payloads that either orbit the earth (satellites) or transit through space (warheads); the signals, wavelengths, or frequencies over which information is transmitted to the user; and the ground components, stations, and analysts or end users that receive and make use of the information. In the simplest sense, it is the end users that determine peaceful or military use, with the peaceful category divided into two groups—civil science and commercial. As far as the other three components, it is extremely difficult to differentiate peaceful from military use.

For some time, however, access has been implicitly differentiated. Rockets carry peaceful payloads into earth orbit or beyond, and ballistic missiles carry military payloads—warheads—through space to terrestrial targets, even though the basic technologies are the same. This differentiation is functional, and the term *ballistic missile* stems from the physics of a ballistic trajectory. The term *missile* can also be traced back to archery—arrows. Regardless, rockets also carry military payloads into orbit, the military employs civil and commercial rockets to place its payloads into orbit, and there is a close working relationship between all three. In most countries, including the former Soviet Union and China, the military and civil programs are closely intertwined. One good example in the United States is the space shuttle. It is a joint U.S. Department of Defense—NASA effort, manned by military and civilian personnel, and carries civil and military payloads into orbit. It is, in effect, a dual-use vehicle for access to space.

Similarly, there are devoted military payloads, or satellites; however, the military also employs commercial satellites, and the civil-commercial community employs military satellites. For example, the U.S. military possesses a dedicated military satellite communications system—MILSATCOM—and also relies significantly on commercial satellite communication systems. In fact, most nations' militaries rely exclusively upon commercial systems since they lack the resources for a dedicated military system. The roots of the Global Positioning System (GPS)—NAVSTAR—are found in the U.S. Navy's Transit navigation satellite. Today, GPS, managed by the U.S. Air Force, is an international public good, which can be used by anybody with the appropriate receiver. Even so, the satellites broadcast two different signals—a more accurate encrypted signal for military use (P-code) and a less accurate open signal for general use (C-code).

Another dimension of the intertwined nature of peaceful and military space is space surveillance, which employs ground-based and, in the near future, space-based sensors to track objects in space. The U.S. Space Surveillance Network (SSN) serves both peaceful and military functions. In tracking thousands of objects in space—satellites and man-made and natural debris trapped on-orbit—valuable information is obtained, for example, to ensure that the space shuttle is not threatened by debris on-orbit. It also serves to identify relatively safe orbits and launch windows for satellites, and it provides valuable warning of deorbiting satellites and meteorites. Militarily, the SSN provides a basic capability to monitor potential threats to military satellites and ideally distinguish between satellite failures as a function of natural causes and failures due to purposeful military action. It also is vital in the process of providing early warning of ballistic missile attack by ensuring that non-military activities (e.g., deorbiting satellites) are not confused with a nuclear attack on North America. Importantly, the SSN is managed by the U.S. military, and its information is analyzed and accessed by the binational Canada-U.S. North American Aerospace Defense Command (NORAD) in Colorado Springs, Colorado.

As the standard bearers for entire ways of life, the cold war superpowers could hardly avoid mixing civilian space achievements and national security requirements. The Soviet Union's early space endeavors, the first satellite (Sputnik) and the first man in space (Yuri Gagarin), were not only great technological achievements, but also powerful propaganda tools. These raised concerns in the United States and among its allies about Western technological prowess. Today, the race to the moon is portrayed as a great scientific feat and a peaceful outlet for technological creativity. In reality, it was the national prestige effects and the potential political consequences of losing the race that drove the superpowers to invest massive resources and sacrifice many lives to win the race. In fact, the Soviet Union spent many years after the triumph of *Apollo 11* denying the existence of its manned lunar program. In an effort to downplay the significance of the U.S. victory and draw attention away from its lunar program, the Soviet Union began to emphasize its space station efforts and successes. Later on, it was revealed that these space stations had largely been covers for extended military missions. Regardless, the efforts made by the Soviet Union to cover-up their unsuccessful bid for the moon speaks to how important manned space exploration was to the cold war rivalry. Indeed, years after the U.S. victory in the cold war, it is hard to believe that early Soviet successes had raised doubts in the West and the United States about our way of life.

National technological prestige was at the heart of the cold war space race between the Soviet Union and the United States. Since then, space technology has remained a measuring stick of a nation's ability to compete on a global scale, not dissimilar in many ways to the prestige associated with building battleships more than a hundred years ago. Many states, seeking to elevate their status on the world stage, are pursuing space capabilities. India, for example, alongside its nuclear weapons program, has developed an impressive set of space capabilities. Most recently, Iran joined the space-faring club with its first space launch, and North Korea's most recent long-range missile test was touted as an attempt to launch a satellite into orbit. Space has remained a medium of global political, technological, and military competition, and with it the continued blurring of military, civil-science, and commercial space.

This blurring has led to a further conceptual distinction between militarization and weaponization. The former reflects the reality that outer space has always been a domain of military use, even though there are some states that continue to call for a ban on all military activity. The latter represents one specific form of military activity, generally understood along a spectrum from the capacity to destroy satellites on-orbit (antisatellite weapons or ASATs) to the placement of weaponized satellites on-orbit. Except for the development and testing of a co-orbital antisatellite satellite system by the Soviet Union in the 1970s, no weapons have been placed on permanent orbit to date. Also, the 1967 Outer Space Treaty explicitly prohibits the deployment of weapons of mass destruction (biological, chemical, and nuclear) on-orbit. However, there is no agreed legal prohibition on conventional weapons, and the possibility that the United States or one of the other space-faring nations might undertake this step in the near future is of major concern and a topic of debate, especially within the United States.

The absence of weaponized satellites has led some to define space as a sanctuary from war.[2] Terrestrial warfare has not been extended into space. However, there is no international legal prohibition that supports this. Moreover, both the United States and the Soviet Union during the cold war developed and deployed (and after the cold war stood down) antisatellite weapon systems. Most recently, China demonstrated the capability to extend war into space with the January 2007 test of a ballistic missile in an antisatellite role that destroyed one of its own satellites in low earth orbit (LEO).[3] Indeed, any country capable of a space launch or possessing a ballistic missile capability has the basic ability to strike at satellites in predictable orbits and thus extend war into space. The United States missile defense capability is also a potential antisatellite system, as most recently demonstrated when a U.S. naval-based missile defense interceptor destroyed a failed deorbiting U.S. intelligence satellite.[4]

While most attention is placed on issues surrounding the employment of kinetic weapons in space, or from space in the future, weaponization also includes potential directed energy weapons and a range of passive mechanisms. As the key value of satellites is to acquire and transmit information, this information is vulnerable to electronic warfare. There are a variety of ways such information can be disrupted or degraded without destroying satellites. These include blinding satellites through the use of lasers, as demonstrated in the U.S. MIRACL laser test in 1997[5] and the reported Chinese test in 2006.[6] Analysts also speculate about the possibility of placing a satellite between the earth and another satellite to block its ability to obtain and transmit information. Thus, a proper understanding of war in space also extends beyond the particular obsession with kinetic weaponization.

Beyond the various roles satellites (military, civil, and commercial) play in support of terrestrial military operations (see chapter 2), there is also a direct linkage between dedicated military satellites and peace. This linkage is directly found in the area of deterrence. One of the goals of deterrence is to maintain peace through the use of threats. Simply, an adversary—the deteree—contemplating the use of force to obtain its objective is dissuaded from doing so because the deterer threatens to make the cost of war much greater than any benefits that might be obtained through the use of force. This basic idea has its roots in the cold war strategic nuclear relationship between the United States and the Soviet Union in the condition of mutual assured destruction (MAD). Both were deterred from attacking each other because any attack would result in retaliation. MAD, in effect, was a mutual suicide pact. No matter which side struck first in a strategic nuclear attack, the other side would be able to strike back, resulting in the destruction of both. Of course, the practice and politics of nuclear deterrence during the cold war are much more complicated than presented here.[7] Regardless, all three major uses of satellites in space serve to provide credibility to the threat of retaliation—space-based surveillance providing early warning of a ballistic missile attack; communications enabling the continued exercise of command and control of strategic forces if attacked; and navigation facilitating the targeting of retaliatory forces, especially for submarine-launched ballistic missiles. Indeed, the origins of the military use of outer space is found in the cold war strategic deterrent mission, which would set the stage for future developments.

The Origins of Military Space

World War II saw the first use of outer space for military purposes with the German V-2 attacks in 1944. Of course, in this instance space was only a transit domain for the rockets, and even then, their relatively short-range (basically a medium-range ballistic missile in modern parlance) meant that they barely reached outer space (see chapter 1 on where space begins). Regardless, both the United States and the Soviet Union recognized the potential military significance of rocketry, and this fed their race to capture German scientists and knowledge. These scientists in many ways would become the foundation of their respective military rocket and space programs, and Wernher von Braun, the erstwhile father of the U.S. space program, would become the best known.

As orbital space launch vehicle technology was developed in the 1950s. First the Soviet Union, and then the United States, deployed a satellite, followed by placing a man in orbit. Both began the deployment of their first generation of long-range, liquid-fueled intercontinental ballistic missiles (ICBMs) capable of launching a single nuclear warhead. In the case of the United States, this included the Atlas, which would serve as the foundation for future civil and commercial launch vehicles, and Titan. Their development, which began in the mid-1950s, coincided with a wide variety of military space development programs undertaken by the U.S. Army, Navy, and Air Force.[8] In particular, the Army, having been assigned the missile defense mission in 1958 as a logical extension of its air defense mission, began work on the development of the first generation of missile defense interceptors—Nike X.[9] The Navy worked toward the development of its first generation of submarine-launched ballistic missiles (SLBMs)—Polaris—and the first space-based navigation system—Transit—initially tested in 1960, which would provide valuable information for the U.S. military's NAVSTAR GPS system of today. The Air Force established a series of projects for the antisatellite mission, including investigation into a space-based system under Project SAINT, a space-based satellite reconnaissance program—the Satellite and Missile Observation System (SAMOS)—and the first space-based early warning system—the Missile Defense Alarm System (MIDAS).[10]

In 1960, the Eisenhower administration created the top-secret National Reconnaissance Office (NRO) for the expressed purpose of developing and employing satellites for military intelligence purposes against its cold war adversary, the Soviet Union. Project Corona, the top-secret program to spy on the Soviet Union, which would only become fully public in 1995, employed a camera-equipped satellite in LEO to photograph Soviet military installations. Once the photographs were taken, the satellite was programmed to release the film canister over ocean, where a U.S. tracking aircraft captured the parachute-equipped canister in midair as it slowly descended. While the time lag between taking the pictures, recovery of the canister, shipping of the film to the NRO, and analysis was significant, the results provided highly useful intelligence of Soviet strategic military capabilities. Moreover, the Soviet Union lacked the capabilities to intercept the satellites. Prior to the development of orbiting surveillance satellites, the United States relied upon high-altitude U-2 intelligence gathering flights over the Soviet Union. These had become vulnerable to Soviet air defenses, as most clearly shown with the shooting down of a U-2 and the trial of its pilot, Gary Powers, for espionage in 1960.

These various developments, and particularly the use of space for intelligence gathering, raised the issue of the status of outer space in international law. The United Nations Committee on the Peaceful Use of Outer Space, established in 1959, became the primary negotiating venue for the subsequent Outer Space Treaty (OST) of 1967.[11] In reality, however, the treaty was largely the product of U.S. strategic military interests. The Eisenhower administration saw space as a vital means to obtain intelligence of Soviet military activities. Orbiting surveillance satellites provided a means for the United States to look into the closed society of the Soviet Union. Although surveillance satellites and the ability to intercept satellites (antisatellite weapons) were still in their infancy, unless some agreement was reached to permit the overflying of national territory by satellites in space, it would be just a matter of time before satellites would potentially become as vulnerable as the U-2 spy planes. The solution was to enshrine outer space in a manner that would permit the right of free passage, and the way to do so was to draw directly upon the treatment of the high seas or international waters in international law.[12] Although the Soviet Union for some time opposed this solution, arguing that it amounted to legalized espionage, it had largely set the stage for this development with Sputnik and the fact that the United States issued no objection to its orbit. Regardless, the Soviet leadership also came to recognize that it too could benefit from the ability to fly over the United States without fear of attack. The net result was the OST, which provided for the right of free passage to all states and, in subsequent international agreements, defined satellites as national property, no different from flagged ships on the high seas. As such, an attack on another nation's satellite can be considered an act of war.

The preamble of the OST states: "The exploration and use of outer space, including the moon and other celestial bodies, shall be carried out for the benefit and in the interests of all countries, irrespective of their degree of economic or scientific development, and shall be the province of all mankind."[13] While this idealized statement has been interpreted by many as prohibiting the military use of space, the reality is that space law was a function of the dominant strategic interests of the Soviet Union and the United States. This also extended to Article IV of the treaty, which prohibits the stationing of weapons of mass destruction on-orbit. This article reflects these strategic interests in two ways. First, it implicitly legalizes the transit of ballistic missile nuclear warheads through space, and this fed the distinction between orbital launch vehicles and ballistic missiles. Second, both superpowers recognized the dangers presented by weapons of mass destruction on orbit, which could strike at terrestrial targets with little, if any, warning. Importantly, the treaty did not explicitly define space, and initial legal questions would emerge, but not be resolved, out of concerns related to the Soviet interest in a fractional orbital bombardment system (FOB). Such a system entailed a warhead launched from the Soviet Union on a southern trajectory around the South Pole, reaching partial (thus fractional) orbit, on its way to targets in North America. To date, no explicit legal definition of space exists. The best one can infer that space begins at the altitude a satellite can complete a single orbit, and this varies, as discussed in chapter 1.

While superpower military-strategic interests were dominant in negotiations leading to the OST, the physics of space also played a major role. The realities of orbital dynamics (see chapter 1) made the idea of extending sovereign airspace into outer space very problematic. The laws of Newton and Kepler make it impractical to enforce the concept of sovereign airspace beyond the boundary of space. Except for geostationary orbit (GEO), where a satellite appears to hover over the same piece of territory (in reality, it moves in orbit at a speed approximating the rotation of the earth), satellites in lower orbits move at speeds faster than the earth's rotation—the closer to earth, the faster the satellite. These satellites might pass over a nation many times a day, making the idea of extending airspace up to and beyond GEO somewhat ludicrous. Nonetheless, at least when it comes to GEO, a few equatorial nations have laid claim to the apparent stationary orbits directly above them, most notably through the Bogotá Declaration of 1976 by Ecuador, Colombia, Brazil, Congo, Zaire, Uganda, Kenya, and Indonesia. Their claims, however, have been completely ignored by the international community, not least of all because none of these states have the capability to access these orbits, and their goal has been largely to obtain rent for the use of these orbital slots.[14]

Attempts to expand the legal regime for space, which came to include the 1967 OST, the 1968 Rescue of Astronauts Agreement, the 1969 satellite Registration Convention, the 1976 space Liability Convention, and the unratified 1979 Moon Treaty, have been unsuccessful. Again, this is explained in many ways by strategic military and political considerations. Neither superpower was willing to accept any further possible constraints on its military use of space, and the logical next step would likely have led at a minimum to calls for limitations on the military use of space, including, for example, a ban on antisatellite weapons as would be proposed by Canadian Prime Minister Pierre Trudeau at the Second United Nations Special Session on Disarmament on June 18, 1982. Both the Soviet Union and the United States had active antisatellite weapons development programs. Of course for political reasons, both participated in the Conference of Disarmament's Committee on the Prevention of an Arms Race in Outer Space but had no strategic interests to achieve any meaningful results. Moreover, by 1979 the international political climate had changed dramatically. The era of relatively good relations between Moscow and Washington, known as détente, which had facilitated the creation of the limited legal regime, gave way to confrontation—the so-called second cold war, and with it, the collapse of arms-control negotiations in 1983 over the question of intermediate-range nuclear forces in Europe.

Even after the end of the cold war, when political conditions arguably were ripe for further developments concerning the outer-space legal regime, nothing of major significance has occurred. Instead, the United States since the Clinton administration has annually blocked the establishment of a work agenda for the Committee on the Prevention of an Arms Race in Outer Space on the grounds that there is no arms race in outer space, and thus there is nothing to discuss. Some small steps were taken with the 2002 Hague Code of Conduct on ballistic missile proliferation, which concerned launch notification, and a similar bilateral Russia-U.S. agreement, but both remain voluntary with no enforcement procedures.[15]

Throughout the cold war, the initial technologies enabling the exploitation of space were a product of superpower strategic interests, as reflected in their military research and development programs. As noted above, both worked on developing a basic antisatellite system, with the Soviet Union developing a system to send a killer satellite into orbit to destroy a satellite by detonating close to the target and creating a high-speed debris field for the target to collide with. The United States developed an air-launched kinetic-kill (direct collision) weapon launched from a high-altitude F-15 fighter. In addition, both experimented with exotic interception technologies, including first-generation lasers and directed energy weapons. Importantly, their antisatellite programs were directed toward intercepting targets in LEO. Both eschewed the development of higher altitude (GEO) antisatellite weapons, partially because of strategic nuclear concerns. GEO housed the critical infrared early warning and strategic military communication satellites, and the development and testing of an antisatellite weapon against these targets had major implications for the stability of the strategic nuclear relationship between them. The capability to strike at these targets, notwithstanding technological barriers and costs, meant that one could blind its opponent to a strategic nuclear first strike and undermine its retaliatory capabilities after such an attack. Such a capability was a harbinger of a first-strike capability and thus generated fears of a possible surprise attack and incentives to establish a dangerous launch-on-warning posture, instead of a more stable launch-under-attack posture, because of the 30 or so minutes of warning provided by the infrared satellites, and the likely loss of command and control of retaliatory forces. Importantly, their existing antisatellite systems were deactivated with the end of the cold war.

Both the Soviet Union and the United States were also working on missile defense systems, known as antiballistic missiles (ABMs) during this period. Both developed a high-altitude nuclear interception capability designed to detonate in the upper reaches of the atmosphere when the incoming warhead was reentering the atmosphere during what is known as the terminal phase. The Soviet Union deployed its ABM system, known as Galosh, around Moscow in 1970, which remains operational today. The United States deployed its ABM system, known as Safeguard, in North Dakota in 1975 to defend the Grand Forks ICBM field. Safeguard was canceled shortly thereafter, and the system disbanded. Even so, the United States continued work on missile defense beginning with SDI, launched by President Reagan in 1983, followed by President George H. Bush's Global Protection Against Limited Strikes (GPALS) proposal in 1991, President Clinton's National Missile Defense development program announced in 1996, and finally, George W. Bush's midcourse phase system, which is deployed in Alaska and California, and became operational in 2004. At the same time, the United States also had developed tactical missile defense capabilities—the Patriot first used in the Gulf War—and land- and sea-based theater systems. The United States has also been developing an air-based laser system.

Although missile defense capabilities are designed to shoot down warheads launched by a ballistic missile, such capabilities also can potentially reach into outer space to target satellites, as demonstrated by the 2008 U.S. use of its sea-based theater system to destroy a failed deorbiting intelligence satellite. Importantly, the U.S. ground-based system and the theater systems are designed to intercept warheads during their transit through outer space, what is known as the midcourse phase, by colliding (kinetic kill) with the target. As such, they are also capable of striking at satellites in LEO. In other words, missile defense can serve two functions—antiballistic missile and antisatellite.

The respective military ballistic missile, antisatellite, and missile defense programs were all driven by the demands of the superpower strategic relationship. These, in turn, led to the development of supporting space-based assets.[16] In 1970, the initial early warning function of MIDAS was replaced by the Defense Support Program (DSP), consisting of a constellation of infrared satellites in GEO designed to identify ballistic missile/rocket launches on the basis of the rocket's heat signature at launch and provide early warning through NORAD of a nuclear attack. The Soviet Union developed a similar system, Oko (Eye) satellites, with the first launch in 1975, and the full constellation of nine satellites operational in 1987. Following the collapse of the Soviet Union, significant gaps appeared in Russia's early warning system due to lack of funding for replacement satellites.

To support strategic command and control, the first operational constellation of communication satellites in GEO, known as the Initial Defense Satellite Communications System (IDSCS), was deployed in 1966. These have been followed by the Defense Satellite Communications Systems (DSCS) series and in the early 1990s by MILSTAR, among other military satellite communication systems. The Soviet Union also developed GEO communications Geizer satellites in 1982. Finally, as noted above, the U.S. space-based navigation system—Transit—for updating the inertial navigation system of Polaris was first used successfully in 1964 and opened to civil users in 1967, with the NAVSTAR development program established in 1973 and the first satellite launched in 1983. The Soviet navigation system, now known as Global Navigation Satellite System (GLONASS) began in 1967 with the Tsyklon (cyclone) series of satellites. The first GLONASS satellite was launched in 1982. Alongside these basic military observation, communication, and navigation space systems, both the United States and the Soviet Union developed and deployed a range of electronic intelligence, geodetic, meteorological, and reconnaissance satellite systems to support military requirements.

The development of these space systems also played a major role in supporting arms-control agreements in the 1970s and beyond. The nature of the relationship between the United States and the Soviet Union made it unlikely that either would agree to physical on-site inspection of their respective strategic nuclear capabilities. Space systems provided the means to verify compliance with the negotiated arms-control agreements, such as the Strategic Arm Limitation Treaty (1972) and the Anti-Ballistic Missile Treaty (1972). Both agreements enshrined national technical means as the key mechanism for verifying national compliance, and these means were space-based. This also served to reinforce the OST's treatment of space as providing freedom of passage.

The emphasis on cold war strategic military requirements would slowly give way to new operational functions in support of deployed forces overseas. This was a function of the end of the cold war and the development of new more advanced

The Evolution of Military Space

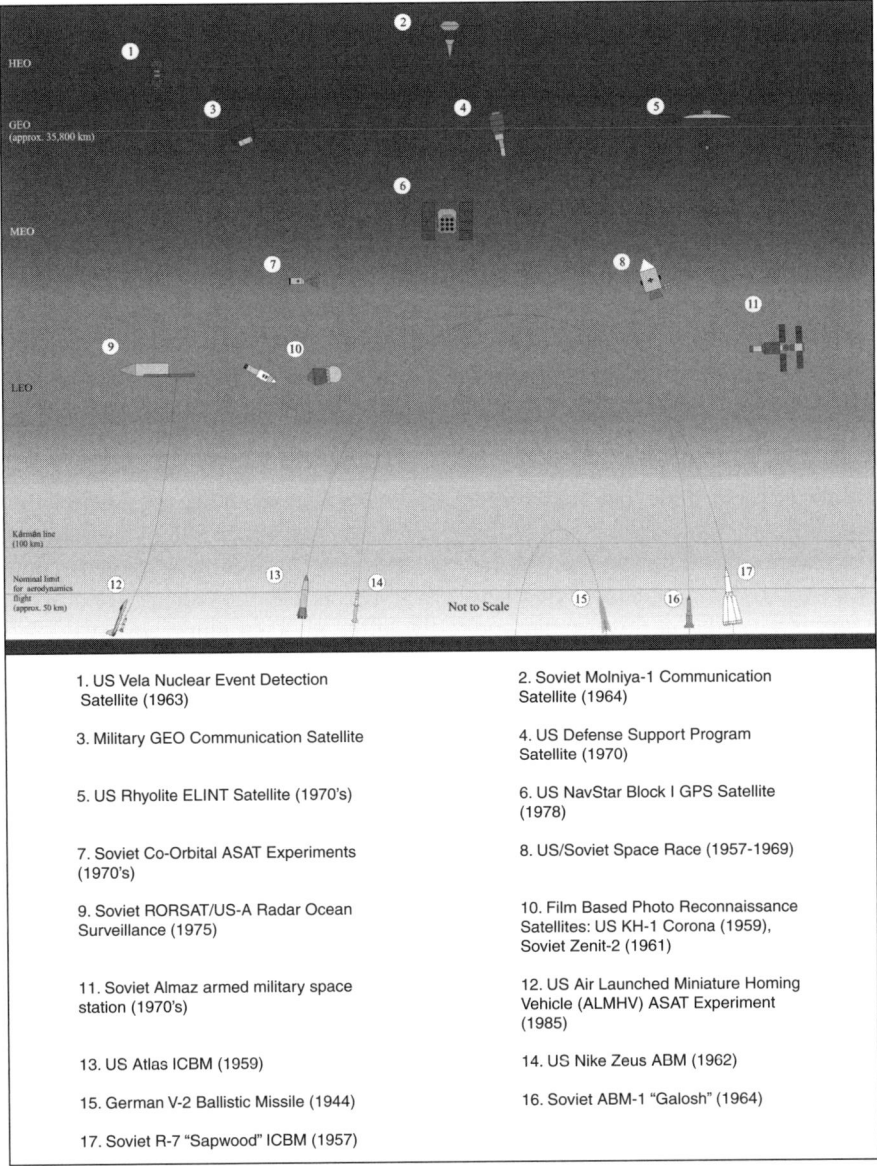

Figure 1. Past Military Space

technologies. Of course, strategic requirements have not been ignored and are likely to increase as further technological advances occur. Moreover, with the new generation of space-faring nations—China and India in particular—with their space programs following a similar pattern as the Soviet Union and the United States, significant emphasis is likely to return to strategic demands.

Conclusion

There is a tendency among many analysts and students to see technology as the key driver in the exploration and exploitation of outer space. This is particularly true in the United States, which has a technologically driven society in many respects. While technology is certainly vital, technologies need supporters and advocates in order to acquire the resources and capital to bring them out of the laboratory and into functional reality. In the case of outer space, the key supporters and advocates in this regard have been found in the military-security realm as driven by the demands of the cold war. The first space age, as some observers call it, is largely a story about strategic cold war military requirements. Certainly, human curiosity and adventurism, which fed the great age of exploration of the New World, would have inevitably led to space exploration, regardless of the cold war. Nonetheless, the manner and pace in which space was explored and exploited during this first age has been a product of military requirements.

Today, the commercial use of outer space has overtaken military use, notwithstanding the significant resources that governments and their militaries invest in commercial space for defense and security reasons. The second space age is thus more complicated in many ways than the first. Regardless, the military use of space remains a significant driver today and in the future in determining how space will unfold. Understanding space thus requires an understanding not only of the medium itself, but also the manner in which the military uses space and indeed thinks about the future of outer space.

To facilitate this understanding, the first chapter provides a background primer of the science and technology related to the harsh and unforgiving environment of outer space. As space is too often understood through the lens of popular science fiction, leading to beliefs inconsistent with reality, it is important to know what is and is not possible given current technology. From this basis, the next chapter looks at the key military developments of what may be called the second military space age—the development of space systems with the purpose of enhancing the application of territorial military force. The next two chapters direct attention to military questions and concerns that have emerged regarding outer space as an independent and unique domain to competition, conflict, and potentially war. The first of these looks at passive measures in this regard. The second looks directly at active measures available and under consideration related to defensive and offensive military kinetic operations in space. Although warfare has not yet extended into outer space, and it remains a relatively pristine environment, the possibility that conflict and war may eventually extend into space cannot be ignored.

Notes

1. Office of Science and Technology Policy, *U.S. National Space Policy*, August 31, 2006, p. 1, http://www.ostp.gov/galleries/default-file/Unclassified%20National%20Space%20Policy%20—%20FINAL.pdf.

2. Bruce DeBlois, "Space Sanctuary: A Viable National Strategy," *Aerospace Power Journal* (Winter 1998), http://www.airpower.maxwell.af.mil/airchronicles/apj/apj98/win98/deblois.html.

3. Craig Covault, "Chinese Test Anti-Satellite Weapon," *Aviation Week & Space Technology,* January 17, 2007, http://www.aviationweek.com/aw/generic/story_generic.jsp?channel=awst&id=news/CHI01177.xml. See also, Shirley Kan, *China's Anti-Satellite Weapon Test,* Congressional Research Service Report to Congress, April 23, 2007.

4. Department of Defense, press release, http://www.defenselink.mil/releases/release.apx?releaseid=11704.

5. Frank Sietzen, "Laser Hits Orbiting Satellite in Beam Test," *Space Daily,* October 20, 1997, http://www.spacedaily.com/news/laser-97a.html.

6. Warren Ferster and Colin Clark, "NRO Confirms Chinese Laser Test Illuminated U.S. Spacecraft," *Space News,* October 6, 2006, http://www.space.com/spacenews/archive06/chinalaser_1002.html.

7. For the best overview, see Freidman, *The Evolution of Nuclear Strategy,* 2nd ed. (London: MacMillan, 1989).

8. Paul B. Stares, *The Militarization of Space: US Policy, 1945–1984* (Ithaca, NY: Cornell University Press, 1985).

9. Gregory Herken, *Counsels of War* (New York: Oxford University Press, 1987), 187.

10. See also Bill Rose, *Military Space Technology* (Hersham, Surrey, UK: Midland Publishing, 2008).

11. See United Nations Office for Outer Space Affairs, *The United Nations Committee on the Peaceful Uses of Outer Space,* http://www.oosa.unvienna.org/oosa/COPUOS/copuos.html.

12. M. J. Peterson, "The Use of Analogies in Developing Outer Space Law," *International Organization* 51 (1997): 2.

13. U.S. Department of State, "Treaty on Principles Governing the Activities of States in the Exploration and Use of Outer Space, Including the Moon and Other Celestial Bodies," January 27, 1967, http://www.state.gov/t/ac/trt/5181.htm.

14. Everett C. Dolman, *Astropolitik* (Portland, OR: Frank Cass, 2002), 134.

15. Center for Non-Proliferation Studies, *Hague Code of Conduct Against Ballistic Missile Proliferation,* http://cns.miis.edu/inventory/pdfs/icoc.pdf.

16. A useful source of national military programs is available from Jane's *Space Directory,* various editions, London.

CHAPTER 1

The Scientific and Technological Foundations of Military Space

To understand military space as a realm of the possible, it is necessary to understand some of the realities that govern it. The physical properties of the space environment define just what can, and cannot, be done in space technology, space policy, and space strategy. Ultimately, space is a domain of high energy. Great amounts of energy must be harnessed to get into space. Great amounts of energy are involved with staying, maneuvering, and operating in space. The energy requirements demanded of the space environment translate into the large financial sums for which space policy is known.

In the late 1950s, the U.S. Air Force coined the term *aerospace,* suggesting a commonality between air and space. In so doing, it laid claim to space as a logical air force responsibility. Unfortunately, the term does not capture how energy demands, technologies, and even international relations differ between air and space. Moreover, the division between air and space is somewhat misleading, and it is more useful to conceptualize the area above the earth's surface as consisting of three parts—the lower atmosphere where conventional aircraft may operate; outer space where satellites orbit; and an in-between region where aircraft cannot fly and satellites cannot orbit, often referred to as near-space, or suborbital space.

Naturally, these three operating areas are not the only method of dividing up the environment above the surface of the earth. Science often finds it useful to divide the atmosphere into the troposphere, stratosphere, mesosphere, thermosphere, and exosphere layers before leaving the final fringes of the earth's atmosphere. These layers are largely defined by the different physical characteristics that each discernable layer possesses. However, with respect to spaceflight basics this refined model of the atmosphere is somewhat unwieldy. The upper limits of the exosphere, for instance, extend to an altitude of approximately 10,000 kilometers, which is significantly higher than the orbits of most crewed spacecraft such as the International Space Station (ISS), which ranges from 370 to 460 kilometers[1] along its nominal orbit. For now, it is useful to concentrate on the basic three parts of the environment

The Scientific and Technological Foundations of Military Space

above the surface of the earth, which largely informs all basic discussions of military application.

Demarcating Outer Space

The lowest levels of the atmosphere are the domain of conventional aviation and airpower. At these altitudes, the lower atmosphere contains sufficient oxygen for a common gas-turbine jet-aircraft engine to operate. The air is also relatively thick, allowing for the easy generation of aerodynamic lift by moving an appropriately shaped object, such as an airfoil or wing, through the atmosphere at speed. The thickness of the atmosphere, which enables powered flight, also imposes limits on flight. Heating caused by the friction of a plane traveling through the atmosphere at high speeds limits most conventional aircraft to just over Mach 2, or twice the speed of sound. Prolonged operations above Mach 2 generate heat loads that require materials and technologies more expensive than aircraft aluminum and conventional aircraft design.

Another way to think of the upper boundary of conventional flight is to consider the technology and performance of the Blackbird family of surveillance aircraft produced by Lockheed's legendary Skunk Works. Though there are faster and higher flying air vehicles, the Blackbird is a useful example because it was an operational, not experimental, aircraft that could take off and land under its own power. The Blackbird lineage of aircraft first entered service with the Central Intelligence Agency in the form of the A-12, and with the U.S. Air Force as the SR-71, both tasked with intelligence gathering. In addition to the two reconnaissance versions that entered service, a prototype interceptor (YF-12) and high-speed drone launch platform (M-12/D-21) also made it into the air before these branches of Blackbird development were cancelled. The original design studies started in the late 1950s and culminated in a family of aircrafts designed for sustained cruise at Mach +3 (over three times the speed of sound). The Blackbird achieved its record performance through the use of air-breathing turbo-ramjet engines (Pratt & Whitney J58) burning a hydrocarbon fuel.

Sustained flight at Mach 3 is not simply a matter of propulsion, but also of surviving the experience. The extreme heat generated by the Blackbird's flight regime led to technical challenges in materials, construction, flight control, fuels, and ultimately protecting the mission payload and crew. For all the effort that went into making the Blackbird, a technological masterpiece with world-beating performance, it is still only an aircraft.[2] Operationally, the SR-71 operated at over 3,200 kilometers/hour at altitudes over 25,000 meters.[3] These figures, while impressive, are representative of the upper boundaries where conventional aviation can be sustained.

Climbing above the altitudes where even the Blackbird can comfortably fly, the atmosphere continues to get thinner, making it progressively more difficult to generate aerodynamic lift and even more difficult to support air-breathing engines. Yet, the atmosphere at these high altitudes is still too thick to allow unpowered orbiting until significantly more altitude is achieved. In addition to aerodynamic drag inhibiting the speeds necessary for orbit are the more destructive effects of very high Mach numbers through relatively dense amounts of atmosphere. Flight though the atmosphere at very high speed generates a great deal of heat from the compression of the air and, from friction with the air. As a result, material from space that ploughs through the upper end of this band of atmosphere has a tendency to break apart from the resultant aerodynamic stresses and the heat generated. A meteor is the visible result of the heat generated by an object passing rapidly through the fringes of the upper atmosphere of near-space.

Due to the difficulty in loitering in this region of the atmosphere, the near-space or suborbital region of the vertical domain is relatively unexploited. Aircraft cannot reach this region, and spacecraft quickly pass through this region to escape the limitations the atmosphere imposes even at these extreme altitudes. Aside from traversing near-space to get somewhere else, high-altitude balloons and research sounding rockets represent the primary examples of human activity conducted at these altitudes today. Advances in sustained flight in near-space, large unmanned balloons and aircraft, may allow near-space systems in the future to fulfill many of the tasks now done by satellites.

At 100 kilometers above the surface of the earth is the Kármán line—one of the more common demarcations for where space begins. Theodore von Kármán calculated that at around 100 kilometers of altitude, the speed needed to generate enough aerodynamic lift to support an object was higher than the velocity of an object in orbit at that altitude. As a result, the object at orbital velocity would be able to avoid hitting the earth without the need for lift generated by aerodynamics or propulsion. In the real world, however, the uneven nature of the atmosphere and the different lift properties of different wings and lifting bodies lead to variation in the actual altitude for orbital velocity. Nonetheless, 100 kilometers was seen as close enough to the calculated value and is easy enough to remember, even though maintaining a circular orbit of 100 kilometers would require active propulsion to maintain sufficient speed to stay in orbit as there remains enough atmosphere for the rapid decay of a space craft altitude and its reentry into the lower atmosphere. Operationally, the lowest of stable orbits are several dozen kilometers above the Kármán line.

While the Kármán line is for the most part defined as the international standard for the boundary to outer space, this is not legally embedded in the 1967 Outer Space Treaty (OST), and the United States uses no official demarcation for space. Instead, the criteria for spaceflight have been dependent on context and/or the agency involved. For example, the altitude criterion to qualify for astronaut wings historically has varied depending on whether one is flying for NASA, the U.S. Air Force, or as a civilian for private interests. At present in the United States, to earn astronaut wings one must achieve an altitude of at least 50 miles (80 km).[4]

Reaching the altitude to qualify for outer space does not mean an object is able to stay in space for any long duration. The LGM-30 Minuteman III intercontinental ballistic missile (ICBM) has a service ceiling of 1,120 kilometers[5] but only stays in space for minutes. Like all ballistic missiles, the maximum altitude of the Minuteman III is a byproduct of its long range. ICBMs converted into launch vehicles boost payloads to initial orbital altitudes well below those achieved by their weapon counterparts. Orbital launch vehicles fly trajectories meant to clear quickly the thicker portion of the earth's atmosphere to where it is possible to achieve orbital velocities, and the key distinction between an ICBM's payload—a warhead—transiting through space and a launch vehicle's payload—a satellite—being placed on-orbit is speed. It requires more speed to place an object on-orbit. Roughly speaking, the payload needs to reach a speed of around Mach 25 to obtain sufficient velocity to enter into orbit, as discussed below.

Powered hovering is one method to extend the duration of an object in space. All that is required is some method of propulsion able to counteract the force of gravity at the altitude desired for hovering. This is essentially the extremely high-altitude version of the direct lift vertical flight concept used by the Harrier family of aircraft. This method of staying in space only works as long as the engine is firing, and compared to achieving orbit, it is impractical with today's technology for any length of time. As the force of gravity decreases with distance from the earth, it is conceivable for a propulsion system with enough endurance to slowly push a spacecraft away from earth until the earth's gravity is no longer of influence. Powered hovering in space has a few applications, such as extending the time a direct-ascent flight profile of a kinetic-kill vehicle is in the path of an oncoming satellite or ballistic missile warhead. However, it is generally better to invest in an orbital launch vehicle if one wants to stay in space for anything longer than the time it takes a weapon to track and hit a target.

When a spacecraft reaches a high enough altitude to minimize aerodynamic considerations and possesses enough velocity in the correct direction, it is in orbit. An object in orbit has, in addition to falling,[6] a sufficient velocity component in a direction perpendicular to a line between it and the center of the celestial body—earth—it is orbiting. This perpendicular velocity in the direction of falling causes the object to continuously miss the celestial body it is orbiting. Baring any further action by onboard propulsion or external forces, the balance of an object's forward velocity and the earth's gravity results in a closed elliptical path around the earth. Orbiting bodies are generally known as satellites, and until the launch of Sputnik 1 on October 4, 1957, the earth only had one—the moon.

Orbits and Satellites

An orbit is described by several characteristics plus some reasonably complex math. Among these are period, inclination, and points along the orbit's elliptical shape.[7] Orbiting is a closed system where energy is not added or subtracted. Only the constant force of gravity and the satellite's own velocity (in a direction tangential to the force of gravity) result in a continuous curved path. The orbital period is the amount of time it takes for a satellite to circle the celestial object—earth in this case. As orbital velocity is a product of altitude, period is related to altitude as well. Generally, the higher the average altitude of the orbit, the longer the period, as is clearly demonstrated by comparing the low-orbiting ISS with the moon in its much higher

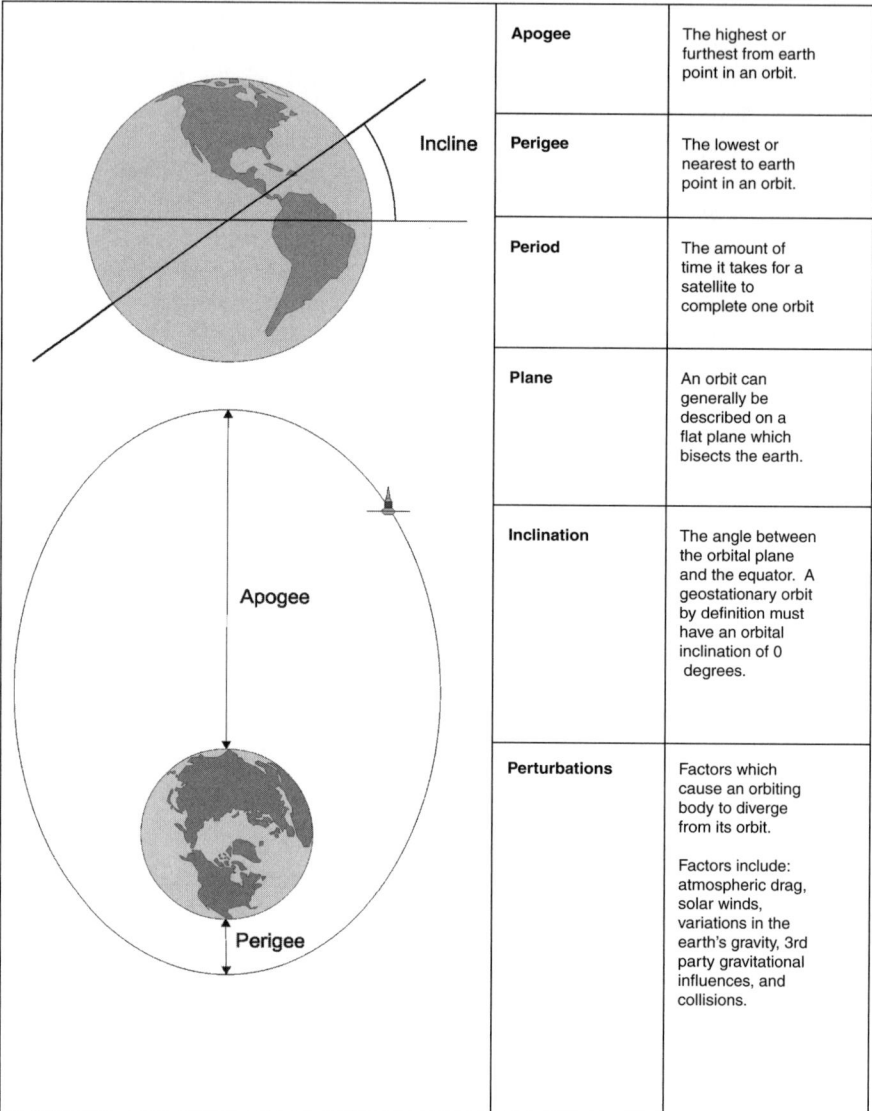

Figure 1.1 Orbital Components

orbit. The space station has an orbital period measured in minutes, compared to the moon's period that is measured in days.

The shape of an orbit forms a flat plane that bisects the earth, which can be oriented in any direction. The inclination is the angle between the flat plane that bisects the earth's equator and the plane of the orbit. Perfectly circular orbits take some effort, leaving most orbits to be elliptical. With respect to earth, an orbit's apogee is its highest point or farthest distance from the earth, and the orbit's perigee is its lowest point. Above all else, an orbit can be described mathematically, which means that a satellite's position above the earth can be calculated from a few variables, and this has significance for a range of military activities, including hiding from surveillance satellites.

Orbital perturbations are factors that cause an orbiting body to depart from the calculated orbit. Perturbations are a function of drag from any atmosphere encountered by the orbit, difference in gravity due to the earth not being the perfect sphere from which basic orbits are calculated, third-party gravitational influences, solar radiation pressure (solar winds), and collisions. These factors can of course be part of a sophisticated orbital calculation if the causes are known. The untimely reentry of Skylab in 1979 is a useful example of an orbital perturbation. High solar activity cause the atmosphere to heat and expand, putting more atmosphere than expected into the path of the unmanned space station, which was in a parking orbit waiting for an early space shuttle flight to boost and service it. This extra air resistance eroded Skylab's velocity enough that it fell from orbit. The nature of Skylab's return to earth resulted in some danger to inhabited parts of earth, with parts of the defunct space station hitting Australia.

Most satellites are not meant to be exposed to the denser layers of the atmosphere or survive atmospheric reentry intact. Nature's usual solution for getting solid objects from space to the surface of the earth is by simply starting off with enough bulk so that after the burning and breaking off of material associated with a trip down through the atmosphere, something survives intact. For human spaceflight, this is known as ablative heat shielding designed to cook off, in order to carry heat away from the object it is protecting (technically this is a more refined system where the heat-shield material changes from a solid directly to a gas). Extinction-event-sized asteroids, such as that thought to have caused the sudden mass extinction of the dinosaurs, are an extreme example of a bulk method of reentry. These asteroids have dimensions estimated in kilometers, thereby ensuring that any mass burned off during atmospheric reentry would be negligible. Meteorites, in contrast, comprise natural debris that survives atmospheric entry, and they usually end up being measured in grams or smaller units of measurement. The earth is constantly bombarded by these much more common and less disastrous bits of rock from space.

Building a structure capable of surviving reentry or designing a mechanism to protect cargo from the harsh temperatures and stresses of reentry is only part of the problem. The correct angle of entry into the denser parts of the atmosphere can lessen the engineering challenges. A shallow enough angle will allow a reentering object to dissipate speed at a controlled rate so that the temperatures generated do not destroy the object. The atmosphere can be thought of as a vehicle's brake pads. Engaging the brake gently converts a vehicle's momentum to heat, slowing the vehicle down. Engaging the brake too hard causes the break pads to overheat (this is more a problem for aircraft brakes than those found on cars).

For any reentry body design, there is a narrow band of angles at which it may safely enter the thicker parts of the earth's atmosphere. Entering at too steep an

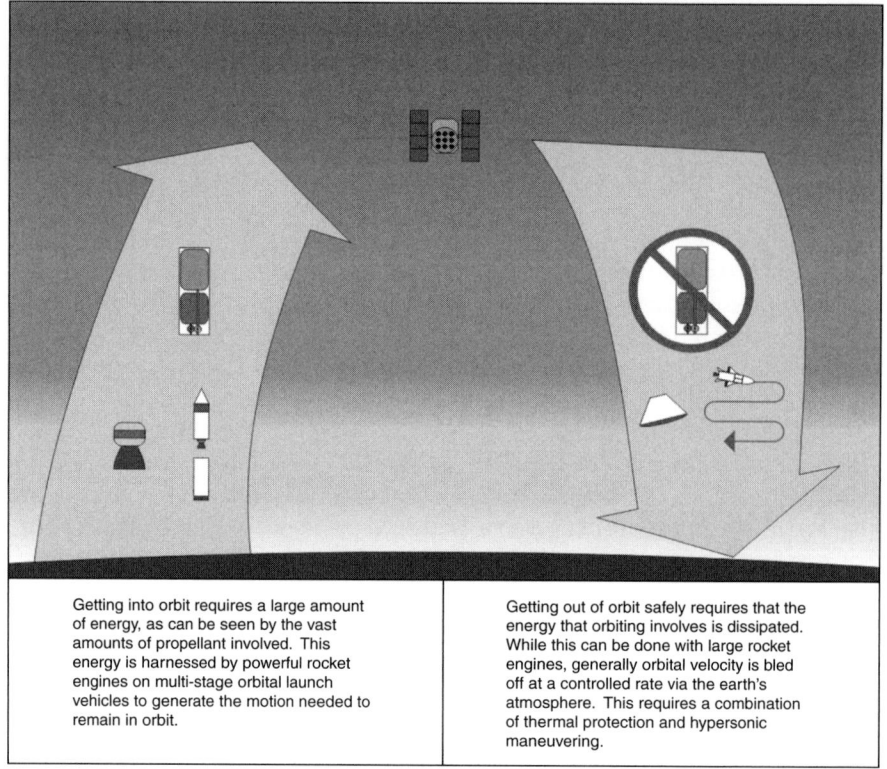

Figure 1.2 In and Out of Orbit

angle will cause the reentry body to exceed its design tolerances with catastrophic results. Too shallow of an atmospheric entry will cause the reentry body to bounce off the atmosphere much in the manner of a stone skipping across a pond. Skipping off the atmosphere in an uncontrolled manner will potentially leave the reentry body stranded in space. Controlled skipping or wave riding off the atmosphere has been proposed as an alternative manner for circling the globe. Since at least Eugene Sänger's "Silverbird" concept for a suborbital "Amerika Bomber" during World War II, there have been proposals for vehicles to skip across the atmosphere, extending range without the extreme altitudes of a ballistic missile trajectory. This is the spaceflight equivalent to terrain masking, or nap of the earth flying, as the speeds involved with this type of suborbital flight and low altitude relative to an ICBM would minimize its exposure above the horizon where it is visible to a ground-based observer.

On the other side of the balance between orbital velocity and gravity, a spacecraft that has more velocity than required for orbiting will break orbit and be flung away from the body it is orbiting. This velocity is known as escape velocity. Notwithstanding the very real threat of impact from some large piece of natural space debris whose path intercepts earth, interplanetary space and beyond are at present of little consequence to national security and power beyond the prestige that deep-space research confers. Understanding the military use of outer space, at least for now, only concerns earth orbit—the spherical volume around the earth that extends outward for thousands of kilometers but is still within the influence of the earth's gravity well.

The lowest earth orbit represents a great investment of energy—much greater than that involved in conventional aviation. Mach 25, twenty-five times the speed of sound at sea level, is the usual benchmark quoted for achieving a reasonably stable low earth orbit (LEO). Speed of this magnitude is practically inconceivable in the lower atmosphere and is only possible at altitudes of around 150 kilometers or more. Though comparisons are made to airpower, transplanting the dogfighting idea of air superiority into space is only a fantasy with current technologies. Aerodynamic support and the lower speeds (kinetic energy/momentums) of aircraft flight share little in common with the careful management of the tremendous energies involved in controlling a spacecraft in orbit. The laws of Newton and Kepler dictate that current spacecraft have less freedom of movement than mammoth warships at sea.

Due to the predictable nature of orbital mechanics, a spacecraft in orbit can in some ways be thought of as an object with a relatively fixed position, like an encampment or a fort. Once a satellite is found and its motion determined, its position at any time (even when not under direct observation) is more or less known. A satellite will remain predictable unless it expends precious propellant to adjust its orbit, and this is only temporary, since once the satellite is found again, its new orbit can be calculated for future reference.

LEO is commonly defined as an earth orbit with an altitude of between 150 and 2,000 kilometers above the earth's surface. For some, achieving LEO represents the most difficult part of exploiting space for any purpose. From LEO, higher orbits can be achieved via carefully calculated engine firings to first establish, and then break out of, transfer orbits. Complete orbital launch vehicles needed to get into a stable parking orbit are certainly more impressive looking than the upper stages and apogee kick motors used to navigate payloads to their mission orbits. LEO has thus far been the destination of most manned space missions (with the U.S. Apollo missions to the moon being the only exception to date). The Iridium constellation of satellites is a relatively recent use for LEO, using 66 active satellites to provide true continuous global communication coverage. Iridium satellite phones, which resemble oversized cellular phones, are quite a bit smaller and more convenient than communication equipment that depends on higher-orbiting satellites. Recent developments in small satellites and small satellite constellations are expected to allow similar global coverage from LEO in roles beyond telecommunications.

Medium earth orbits (MEO) are orbits with an altitude between that of the upper limit of LEO (2,000 km) and below the approximately 35,800 kilometers of a geosynchronous orbit (GSO). The delimitation between LEO and MEO is, like many aspects of space and policy, notional. At least one source puts the lower limit of

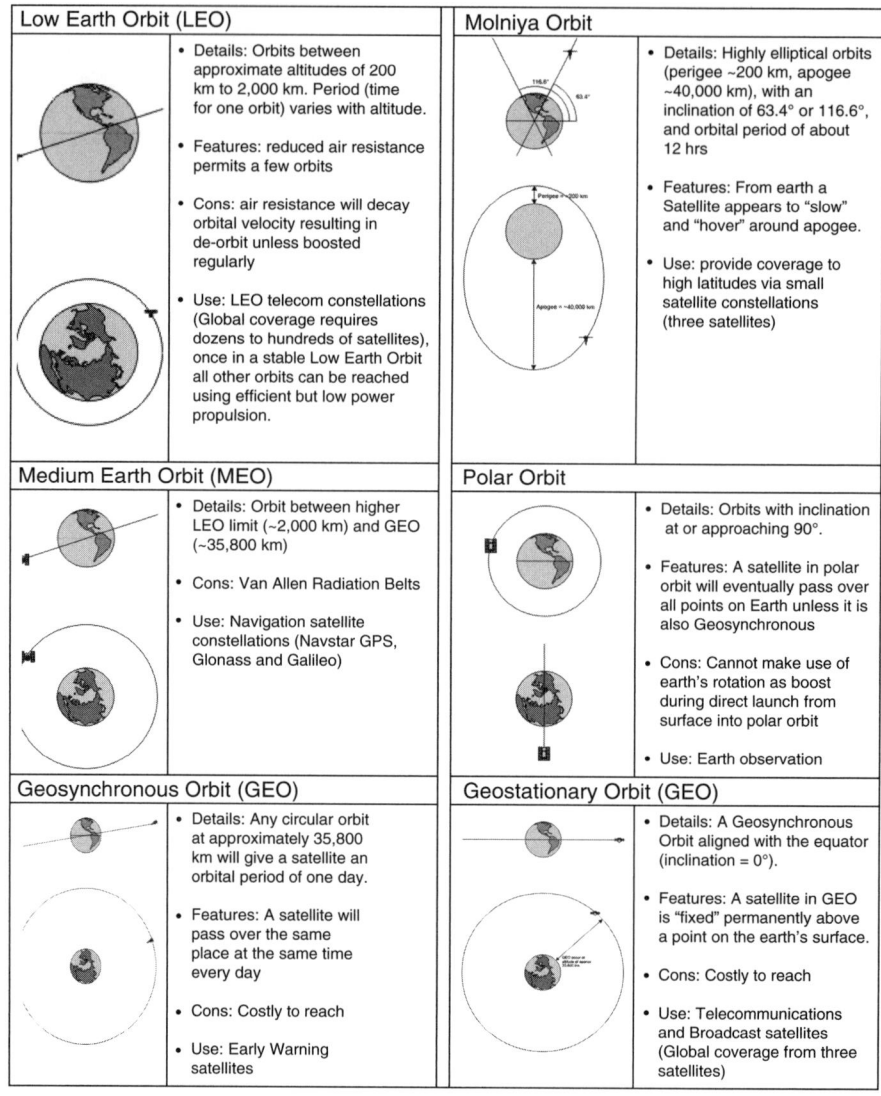

Figure 1.3 Primary Space Orbits

MEO at 300 miles (or ~483 km) above the earth's surface. The lower of the two natural Van Allen radiation belts is sometimes used as the lower limit of MEO, though this radiation belt can extend as low as LEO at lower earth latitudes. Like the atmosphere, the Van Allen belts are dynamic in nature with some variation in shape and consistency due to natural causes. Navigation satellite constellations such as those used in the U.S. NAVSTAR Global Positioning System (GPS) are the more security-relevant satellites found in MEO.

As orbital speed relative to a point on earth's equator reduces with altitude, there exists an orbital altitude where the orbital period matches the period of earth's rotation—a few minutes and seconds less than 24 hours at 23 hours, 56 minutes, and 4 seconds.[8] The altitude at which the orbital period is the same as the earth's rotational period is about 35,800 kilometers. As long as the satellite is orbiting in the same direction as the earth's rotation, the ground track of this satellite will put it over the same spot on the surface at the same time daily. A circular orbit at this altitude is called geosynchronous as it matches the earth's rotation. U.S. Defense Support Program (DSP) satellites, which currently provide early warning of missile launch, are found in GSO.[9] A satellite whose orbital period is half that of the earth's rotational period is in a semisynchronous orbit (SSO) and will consequently pass over the same point on its ground track twice a day at regular times.

A GSO aligned along the earth's equator will have the satellite's movement matched with that of a point on the equator, resulting in the satellite hovering continuously above that point on the equator. This is called a geostationary or geoearth orbit (GEO), although the acronym GEO is sometimes also applied to geosynchronous orbits. It is also sometimes still called the Clarke orbit after author Arthur C. Clarke, who popularized the concept. A satellite in a relatively fixed position above a spot of the earth's surface is of great utility for many roles, including military missions. Three communications satellites in GEO can provide continuous coverage to all points on a band around the earth that encompasses all but the highest of latitudes.

An orbit inclined so it passes over the earth's poles is in polar orbit. A consequence of orbiting with such a high inclination is that it will eventually pass over all points on earth, unless the satellite's altitude is also synchronized with the rotation of the earth. Then, it will pass over the same points on its ground track with clock-like regularity just like any other GSO. The property of passing over every point on earth makes a satellite useful for earth observations ranging from dual-use missions such as producing maps to intelligence gathering.

Timed correctly, highly elliptical orbits can make use of a phenomenon known as *apogee dwell* to provide relatively long-duration coverage of a geographic region during part of its orbit. As a satellite's movement takes it near its apogee in a highly elliptical orbit, its speed along its ground track across the surface of the earth slows. This results in what is called apogee dwell where the satellite seems to dwell for a time over the area of the earth below the orbit's apogee. Conversely, the satellite's ground track will be moving at its fastest around the orbit's perigee. Among those of note are the Molniya orbits used to provide communications coverage to the Soviet Union's, now Russia's, northern territories. Molniya orbits are highly elliptical, swinging between a perigee of about 200 kilometers and an apogee of 40,000 kilometers. Molniya orbits are semisynchronous with orbital periods of approximately half a day and an inclination of 63.4 degrees or 116.6 degrees, which gives the satellite extended coverage on the order of eight hours over high latitudes. This allows coverage to high latitudes with only a small constellation of satellites, which explains their usefulness for northern nations such as Russia to provide satellite communications to remote Arctic communities.

The Hazards of Space

Space is a hostile environment. Without the protection of earth's thick atmosphere, equipment is exposed to intense heating by the sun, extreme cold when in the shadows, and the effects of transitioning between the two extremes (often simultaneously on the same spacecraft). Aside from visible light, spacecraft are directly exposed to other forms of radiation. Without the atmosphere to dampen and restrict attainable velocities, impacts in space can involve incredible energies. Finally, the vacuum of space is lethal to all forms of life. While the pressure difference between sea level and space is only one atmosphere, compared to the many crushing atmospheres of pressure of the deep ocean, vacuum or near vacuum means that there is quite literally nothing for life to work with. Unlike the hostile depths of the ocean, there are none of the substances, particularly sources of oxygen, necessary for biological functions. Out from under the safe blanket of the earth's lower atmosphere, space is an unforgiving environment. As such, space is threatening enough to humans, equipment, and budgets without having to prepare for willful acts of violence.

Objects in stable orbit tend to stay in orbit because without the braking influence of the atmosphere there is nothing to wear down momentum. This includes objects that have left the control of humanity—so-called space garbage. Since the early days of spaceflight, and to this day in science fiction, there have been fears of spacecraft running into grief due to natural space debris—meteor showers being the cliché example. In reality, the primary source of impact hazards come from the remains of satellites that have failed, spent launch vehicle stages, and other artificial debris from the various space programs. These objects are a hazard because as a consequence of human space activity, they occupy the orbits used by operational satellites and manned space missions. Space is vast, but with orbits taking objects around the earth dozens of times per day, it becomes a numbers game for when the next impact will occur.

The speeds involved with orbiting mean that even the smallest piece of space debris has a great potential to cause damage. Kinetic energy is a function of mass and velocity, expressed mathematically as one-half mass times velocity squared ($E_K = \frac{1}{2} m * v^2$). This means that a proportional increase in velocity has a greater effect than an increase in mass. For example, doubling the mass involved only doubles the kinetic energy. Doubling velocity, on the other hand, will result in four times the kinetic energy. Minimal orbital speed, around 25 times the speed of sound, is over eight times the speed available to conventional aviation. Therefore, if one had an object in orbit that was the same mass as the Blackbird aircraft, the orbiting mass would have roughly 64 times the kinetic energy of the Blackbird cruising at Mach 3.

Warfare in space, unless carefully designed to somehow remove the target from orbit (uncontrolled reentry in the atmosphere or more fancifully capturing the target intact by a vehicle capable of surviving reentry), will leave hazardous debris for other spacecraft. Any totally disabled satellite will circle the earth uncontrolled with the potential to impact some other spacecraft. More destructive methods would result in a cloud of debris that would closely follow the original orbit. Smaller debris is difficult to track, and due to the smaller surface area presented to the fringes of the atmosphere, smaller debris may last in orbit longer than an intact dead satellite. The outrage over the 2007 Chinese Anti-Satellite Test (ASAT) was in part over the generation of millions of new pieces of space debris, each with the potential of hitting an operational satellite or crewed spacecraft. More pieces of space garbage mean more opportunities for disastrous impacts. These impacts, in turn, may generate a runaway chain reaction where collisions create more debris, leading to more collisions and more debris, which may result in orbits becoming unusable—what has been termed the Kessler syndrome as discussed in chapter 3.

High-energy subatomic particles or radiation pose another problem to spaceflight. Subatomic particle impacts, unlike those of natural and artificial space debris, constantly occur. Natural radiation sources include the sun and stellar events. Due to the great distances involved, and the fact that radiation generally travels much more slowly than the speed of light, interstellar radiation originates from events often millions of years in the past. Background radiation is the by-product of the birth, life, and death of stars and galaxies. At orbital altitudes, there is not enough atmospheric density to absorb and protect against radiation. Like people, satellite components can only absorb so much radiation before damage is sustained. This accumulated dosage of radiation is a major factor in the life span of a satellite.

Radiation exposure is not uniform in space, but the overall exposure a satellite will face can be estimated. Solar activity research has lead to the important service of solar weather prediction and reporting. On earth, the aurora borealis (or northern lights) are a visible effect of solar activity, particularly from higher latitudes (such as one author's hometown north of the 56th parallel in Manitoba, Canada). Solar activity interference with electronics extends from space down to the surface of the earth causing problems for mundane devices from home high-speed cable modems to entire power grids. Untold numbers of events from the entire history of the universe are responsible for the cosmic ray component of background radiation, and the study of interstellar radiation sources encompasses various disciplines within astronomy. The predicted dosage of radiation a satellite will encounter results in the balance of mission payload verses shielding, or radiation hardening in a satellite's design.

In addition, a satellite's orbit can dramatically raise the amount of radiation exposure. The Van Allen radiation belts represent the prime natural radiation hazard that a satellite may run into due to the topology of space. These belts are two donut-shaped regions of highly charged particle density formed by the earth's magnetic field. Particles enter and leave the Van Allen belts, naturally forming an equilibrium. Normally, the two belts maintain their position and shape relative to the earth. In between the two belts, there is the *slot*—a region where there are lower levels of radiation. The low density of charged particles in the slot is a product of the tendency of the earth's magnetic field's to move particles into either of the two Van Allen belts. Within the belts, however, the higher density of charged particles increases, as does the rate they strike satellites. Consequently, the radiation dosage of a satellite accumulates faster when satellites are in the radiation belts. The mechanisms by which high energy particles leave the Van Allen belts can be overwhelmed by radiation events such as extreme solar activity and high-altitude nuclear detonations,[10] resulting in temporary enhancement of the strength and even the number of radiation belts. These effects can last for months or even years, impacting severely how quickly exposed satellites will fail.

Adding to the cost of surviving space, there is no convenient way to perform maintenance on equipment in space. There is only so much that remotely resetting or power-cycling satellite components can accomplish. Eventually, all computer-based equipment must be serviced or disposed of. There are few examples of on-orbit repair and servicing of satellites, and all that have so far occurred have been rightfully hailed as extraordinary examples of skill, planning, and funding.

A comparison of the airborne laser (ABL) and space-based laser (SBL) programs demonstrates the difference between terrestrial and space-borne operations. Both programs share the common idea of using high-energy lasers for missile defense, and potentially antisatellite (ASAT) missions. The nature of the operating environments has led to some significant technological differences. The ABL program's laser weapon must be able to pass though atmosphere of varying thickness, and includes technology to compensate for distortion, while the equivalent laser envisioned for SBL just needs to deliver high energy through the clarity of space. The fact that the vacuum of space is not an impediment to laser propagation is probably the only area where the nature of the space environment is of benefit to SBL developers. The ABL based on a large Boeing 747 freighter aircraft is expected to be serviced regularly in the comfort of hangars on the ground. In light of existing launch technology, the SBL, as currently envisioned, is to be based on an orbiting satellite, where it will have to function without regular, and likely any, maintenance throughout its operational life span. Furthermore, if deficiencies are discovered in an ABL aircraft after deployment, the system can be adjusted or sent back to the contractor with relative ease on the ground. In contrast, the heroic effort needed to fix the defective Hubble space telescope after it was launched only serves as a warning for the need to get things right the first time when it comes to expensive space platforms like the proposed SBL. This level of engineering perfection only adds to the already considerable problems of space operations.

In a similar vein to maintenance in space is the related subject of construction in space. Due to the limitations of contemporary launch vehicle technology, payloads are severely constrained by mass and physical size. For similar reasons of limited circumference, many satellites are designed to deploy components once reaching orbit, much as one would raise the rigging of a model sailboat or tall ship once it has been stuffed into a bottle. The parts that make self-deployment possible are often only used once and otherwise are dead weight imposing development and payload costs. On the other hand, putting a construction crew into space is quite possibly even more expensive, and certainly more dangerous. The ISS, with the exception of a few Russian launched modules, has been hauled into orbit by the U.S. space shuttle and assembled by a combination of the Canadian-built robot arms mounted on the space station and space walks. Trying to package everything into one launch is often not possible, and tailoring a launch vehicle to one specific mission is a rare and expensive luxury.

With all the dangers just from being in space itself, it becomes difficult to recognize an attack in space. Current limits on space monitoring give rise to fears that the United States, or any other satellite operator, would be unable differentiate satellite failure caused by natural events from that caused by an overt hostile act. For both sides of the space weapons debate, the inability to fully monitor objects in space increases the risk of an accident turning into a diplomatic matter. It should be further recognized that any known dedicated ASAT capability would tend to raise suspicions that the unexpected failure of militarily important satellites was an intentional act of hostility. Under pressure from an international crisis, suspicion easily turns into retaliatory action, leading to a catastrophic chain of events.

Rocket Science

Every launch into orbit requires a great amount of energy. As mentioned before, it is not just a simple matter of going straight up. The forward speed needed to keep a spacecraft from hitting the earth in the continuous free fall of a basic LEO is on the order of 25 times the speed of sound at sea level. Higher orbits require much more speed and thus energy. The great energies needed for satellite placement must be released in a precise manner so that the satellite actually gets to its proper orbit. While there has been much speculation over the ability of technology to scale back the ground crew and bureaucracy (producing leaner launch operations and cost savings along much the same philosophy as found in New Public Management), the energies involved cannot be reduced.

Before accelerating a payload to orbital speeds, a launch vehicle must first overcome the force of gravity. Engineering and science students generally use 9.81 meters per second squared (m/s^2) as the acceleration toward the center of the earth caused by gravity at sea level on earth. To achieve lift, this force of gravity must be first overcome. As an object moves through the air, it encounters drag. The force of drag increases with velocity for a given surface area and constant medium. The propulsion system must therefore have additional capability beyond that needed for achieving orbit to account for the launch vehicle's passage through the atmosphere. Both the force of gravity and air resistance decrease with altitude, though gravity's decrease in force compared to the decrease in air resistance is extremely small. For all intents and purposes, the immediate effects of air resistance at space altitudes are negligible. Gravity, on the other hand, is still a major concern as it is the force that must be cheated by sheer speed to achieve orbit.

Except for a few exotic propulsion schemes (solar/beamed-energy-type sails, electrodynamic tethers), thrust must be produced by throwing material in the opposite direction of flight. Newton's third law states that for every action there is an equal and opposite reaction. Rockets and jets are propelled by the force that occurs in reaction to the expelling of material, or working mass, in the opposite direction. The reactive force is usually referred to as thrust.

A pure rocket is a self-contained system, carrying onboard all the propellant needed for flight. A rocket can be as simple as pressure expelling a fluid; this is the mechanism for cold gas thrusters. For greater levels of thrust, more energy than can be obtained from simple pressure must be harnessed. Chemical rockets, at present the most common form of surface-to-space propulsion, generally use a combustion chemical reaction. Combustion, the chemical combining of fuel and oxidizer, releases energy mostly in the form of heat and supplies reaction mass in the form of the combustion products. The large quantities of heat generated expand and expel the reaction mass out a nozzle, generating thrust. Fuel and oxidizer must be forced into the rocket engine for it to operate, necessitating the use of powerful turbo-pumps or pressurization of thick-walled tanks. Both turbo-pumps and propellant tank pressurization are heavy. Solid rockets mix together solid fuel and oxidizer, often in a particle form held in a binding material. Once ignited, solid fuel rockets cannot be easily shut down or otherwise controlled. As the pure rocket is completely self-contained, launch vehicles that use this type of propulsion (practically all existing ones today) tend to fly trajectories that emphasize quick clearing of the denser parts of the atmosphere with most of the acceleration to orbital velocity occurring at higher altitudes.

In the case of a hydrogen-oxygen propellant rocket motor, liquid hydrogen fuel and liquid oxygen oxidizer, which are both cryogenic (extremely cold) liquids, are combined to generate great amounts of heat to expel the by-product of the reaction, essentially steam. The hydrogen-oxygen reaction, in the slow form of a fuel cell, is now promoted as a potential clean energy source, or more correctly, energy storage medium for operating cars and other vehicles. However an uncontrolled and energetic hydrogen-oxygen reaction was also the mechanism for the destruction of the airship *Hindenburg*, which has been seared into aviation memory. In between is the energetic but controlled hydrogen-oxygen rocket engine. In all three cases, two hydrogen atoms end up bonding with a single oxygen atom to produce a water molecule, releasing energy in the process.

The contemporary launch vehicle sheds a lot of hardware to reach space and, in most cases, requires multiple stages to propel a payload up to orbital velocities. The purpose of an orbital launcher is to give a payload the velocity needed to stay in orbit. In other words, the orbital launcher exists to change the velocity or Delta-V of a payload from its start velocity to its orbital velocity. A rocket by definition is self-contained and therefore has to not only accelerate itself and its payload, but also accelerate whatever propellant happens to be onboard at the time. Konstantin

Tsiolkovsky's rocket equation relates a launch vehicle's capacity to change velocity to engine performance and the vehicle mass both initially and at the end of the velocity change. Based on the performance of a chemical rocket engine measured in exhaust velocity, Tsiolkovsky's rocket equation produces a mass change of around 90 percent or more for a single-stage-rocket-based vehicle to change its velocity from a standstill (ignoring any velocity imparted by the earth's rotation and other sources prior to rocket-borne flight) to orbital velocity. This mass change, or mass fraction, represents the mass of propellant that must be expended to achieve orbit, and the remaining 10 percent or less represents the engine, propellant tanks, structure, flight control, and finally the payload itself.

Staging breaks the problem of reaching orbital velocity into parts. After counteracting gravity and drag, each stage is responsible for only part of the velocity change needed to get up to orbital velocity. Each stage only has to accelerate the payload, the stage itself, its propellant, and any later complete stage or stages with propellant. Tsiolkovsky's rocket equation, led to Tsiolkovsky's concept of using multiple stages, which results in nonpropellant mass fractions being significantly higher than 10 percent. Effectively, the propellant, structure, engines, and other components of later stages are the payload for the initial stage, resulting in the non-first-stage-propellant mass fraction being above half in many launch vehicles.

Rocket engine performance, like that of a car engine, can be measured in many ways. Thrust is a measurement of reactive force and is therefore measured in units of force such as the metric unit of Newtons, or, more commonly for jet engines, pounds of thrust. Tsiolkovsky's rocket equation uses exhaust velocity as the variable to represent engine performance. Exhaust velocity divided by the standard for earth's gravitational acceleration (9.81 m/s^2) gives specific impulse (I_{sp})—a measurement of reaction engine performance in units of seconds (s). Specific impulse, though it can be applied to any reaction engine, is more associated with rocketry than jet engines. The greater the speed of the ejected reaction mass, the greater the efficiency of a rocket converting propellant mass to Delta-V.

The mass that is ejected is important as well. If an engine cannot process enough mass to generate a significant amount of thrust, the vehicle will not be able to defeat gravity. It is for this reason that some electrical engines, such as ion engines, are very efficient because they can produce very high exhaust velocities, resulting in less reaction mass needed to produce a desired Delta-V. Low fuel consumption means that these engines can be run for months on end, resulting in final velocities beyond that possible for chemical rockets. However, ion engines are ineffective for lifting a payload off the earth due to the low levels of thrust generated at any one moment in time. The thrust level of an ion engine is often compared to the weight of a few sheets of paper.

In certain ways, launch vehicle design resembles the art of politics. There are many competing ideas about how best to get into orbit, and, depending on the distribution of needs and goals, all or none of the proposed solutions may be correct. In the real world of competition for research and development funding, more down-to-earth politics are usually involved as well. The result is that real-world launchers, like most everything else in the real world, are based on compromise. Launch vehicle development represents several different balancing acts, most notably between the costs of using what is available and funding incremental improvement versus the costs of chasing risky technologies that promise great payoffs only if the gamble succeeds. Though rocket science is based on precise mathematical calculations, there is an art to how the problem of space launch is solved.

The art of space launch is very expensive. At present, the cost of space launch is often measured in units of thousands of dollars per kilogram (or currency per unit of mass of your choice). A large fraction of this cost is in the ground personnel needed to prepare a launch vehicle for flight. There is also the expense of the development and construction of machines that can generate and survive the tremendous forces associated with space launch, while being as light as possible. Finally, there are the energy requirements themselves. Powerful rocket propellants are relatively expensive. Liquid hydrogen and liquid oxygen, the primary propellants, are cryogenic liquids, which are by nature difficult to handle as they both boil off at temperatures hundreds of degrees below zero. The key to greater access to space is in finding a way to lower the cost of getting into space.

For various reasons, existing heavy launch vehicles do not follow the engineering principle of keeping things simple. Describing the U.S. space shuttle as the most complex flying machine in existence was once a point of pride for its supporters. Today, this very complexity is a point of criticism that has proved to be the shuttle's downfall. Many thousands of parts must operate near perfection, which requires a large ground crew to prepare and monitor these parts from launchpad to orbit and back to the earth. Despite this immense effort, accidents have already claimed two space shuttles and their crews. Other launch vehicles have fared no better. Most competing expendable launch vehicles have suffered at least one major launch failure. Oddly enough, the only major exception to the tendency for launch vehicles to explode in flight at least once is the Saturn V heavy lift vehicle that won the moon race for the United States. A counterpoint to the Saturn V's success is the Soviet Union's competing heavy lift moon rocket, the N1, which never managed to leave the launchpad without exploding in spectacular fashion. In the end, space launch is a complex and dangerous technology, although reduction in complexity may improve reliability and safety and thereby reduce costs in the process.

The air-breathing launch vehicles represent another gamble at lowering launch costs by reducing the need to carry onboard oxidizer (air). Air-breathing jet engines such as turbojets, ramjets, supersonic combustion ramjets (scramjets), and air-breathing rocket engines are propelled by the same reaction principle as a rocket. Instead of hauling onboard the large quantities of oxidizer associated with a pure rocket, the air-breathing engines obtain the oxidizer from the atmosphere. As the majority of the oxidizer needed is not onboard, the comparative propellant amount in an air-breathing launch vehicle is much lower than that of a rocket-based vehicle. The reduction of onboard oxidizer reduces launch mass and vehicle size. However, the need to intake and condition air for use by engines, whether by liquefying air for use in a rocket-type engine such as in the British Horizontal Take-Off and Landing (HOTOL) concept of the 1980s or by supersonic combustion ramjets (scramjets), results in heavy engines, which impact deliverable payload.[11] Proponents of pure rocket flight argue that these demands surpass any benefits from not needing to carry large amounts of onboard oxidizer. In addition to the problems of processing atmosphere, pure air-breathing engines cease to function when atmosphere itself does not contain enough oxygen.

Accelerating to orbital velocity at a low enough altitude for a scramjet to work requires a very strong airframe and robust thermal management system. In the 1980s the U.S. X-30 program initially envisioned a scramjet-only flight profile, circulating hydrogen fuel to cool the airframe, recouping energy in the process. Prior to the shelving of the X-30 program, rocket propulsion was added to the scramjet operation, which ceased at progressively lower Mach numbers. In so doing, the lower the velocity of Delta-V provided by scramjet, the lower the technological challenge. However, the less the scramjet did, the more oxidizer was needed, which reduced the practicality of the X-30 reaching orbit. Scramjet research continues to this day, but with much lower near-term speed targets than the 24 times the speed of sound of the original X-30 proposals.

Rockets and jets are both engines that use a form of combustion classified as deflagration. Detonation, a somewhat different and more efficient chemical reaction to the deflagration reaction inside a conventional combustion engine, is now being touted as another near-term way of generating thrust. Detonation is combustion at a supersonic rate (relative to the speed of sound of the explosive medium) where the sustained burning of the explosive is due to a supersonic shockwave and not heat. High explosives when ignited properly burn at rates measured in thousands of meters per second. This is a detonation. In contrast, deflagration as found in most combustion engines (and for that matter gunpowder and other low explosives) is combustion at a rate below the explosive's speed of sound propagated by heat.

Open literature in recent years has highlighted ongoing work to build a pulse detonation engine (PDE) to harness this method of chemical energy release.[12] These engines dispense with the heavy turbo-pumps or thick fuel tanks and pressurization systems necessary to sustain combustion in a conventional rocket and instead inject an explosive fuel-oxidizer mixture in cycles, which generate thrust when detonated. The smooth constant thrust of jet and even rocket engines is replaced by a rapid series of explosions, each pushing the engine along. As jarring as this concept sounds, the detonation reaction represents a more efficient way to extract chemical energy to obtain thrust.

A chemical reaction is not the only means of accelerating a reaction mass. Thermal rockets may use any energy source to heat and expel a working fluid. One controversial proposal has been to use onboard nuclear reactors to directly heat a working fluid to great temperatures to provide higher exhaust velocities or thrust. Though usually confined to deep-space applications, nuclear thermal propulsion has been proposed for atmospheric use in the past. The heat from a nuclear reactor is capable of generating much higher exhaust velocities than a chemical reaction. Flying nuclear reactors within the atmosphere, though once a very promising line of research for long-range aviation and space launch, have been more or less abandoned today due to many real and imagined problems. Among the more legitimate fears are the implications of a nuclear-powered flying machine crash. The stigma associated with nuclear power and flight has extended to the opposition and paranoia expressed over the dangers associated with the use of very small (and safe) radioisotope thermoelectric generator (RTG) power sources for deep-space probes.

Beamed power via microwave or high-energy laser from a stationary ground station has also been proposed and generally does not garner as much opposition as flying nuclear power plants. Flight by lasers has already been demonstrated, where a laser is cycled on and off to superheat air into explosively expanding pulses underneath a scale model of a spacecraft. Essentially, this is a form of a PDE. A beamed-power pulse engine does not require any onboard propellant until reaching high into the atmosphere. The beamed-power propulsion concept has been demonstrated on a small scale by Leik Myrabo of the Rensselaer Polytechnic Institute.

Highlighting the great energy requirements of spaceflight is the combination of nuclear technology and pulse detonation propulsion known as Project Orion. In the 1950s and 1960s, Project Orion proposed using small nuclear bombs to provide thrust. The power released by an Orion-type nuclear pulse engine would have allowed interplanetary expeditions to be launched from the surface of the earth. The impulse from each nuclear bomb would be transmitted to the space ship via the pusher plate and shock absorbers. There remains some enthusiasm for reviving work on the Orion-type engine, though not perhaps for the idea of launching an Orion-powered spacecraft from within the earth's atmosphere. No other existing or near-term technology approaches the performance of the Project Orion nuclear PDE.

In the other direction of rocket development is the use of less exotic propellants. Liquid hydrogen and liquid oxygen have become a popular choice of rocket propellant due to the lack of toxicity in engine exhaust (water) and the provision of high energy content per unit mass. However, to practically use hydrogen and oxygen in a launch vehicle requires the use of their liquid forms, which can only exist at cryogenic temperatures. Cryogenics liquids present production, material, handling, and other problems, all of which have been solved. The problem, however, is the high cost. Up until the early space age, the extremely low temperatures associated with liquid hydrogen were a scientific novelty with few applications. Despite cryogenic liquids handling becoming somewhat commonplace in applications such as cooling medical imaging equipment and liquefying natural gas for transport, it is still a relatively difficult technology to master. Avoiding the problems of cryogenic propellant storage, less-efficient oxidizers such as nitrous oxide (laughing gas) or high-test hydrogen peroxide can be used in combination with easy-to-handle hydrocarbons such as kerosene. Kerosene, diesel, and other proposed hydrocarbon rocket fuels may have less power per unit of mass than hydrogen, but they have a higher energy density. Hydrocarbon-powered stages and launch vehicles having to carry more propellant by mass tend to have smaller fuel tanks, which reduce the overall size and drag a vehicle will experience. The Saturn V used rocket-grade kerosene fuel (RP-1) for its first stage, and liquid hydrogen for its second and third stages, using the fuel best suited to each phase of flight from Florida to the moon and back. Wax-based fuel has also been examined in at least one U.S. Air Force study[13] and demonstrated to the general public on the science-entertainment television program *Mythbusters*.

More simply, solid rocket motors can be used. (Liquid rockets are usually termed *engines*, and solid rockets are termed *motors*.) Solid rocket motors trace their lineage back to the ancient Chinese gunpowder rockets that began rocketry. Contemporary motors are often just aluminum particle fuel with a solid oxidizer held in a rubberlike binding material. There are no heavy pumps and piping for propellants in a solid rocket motor. The fuel tanks and combustion chamber are one in the same. Solid rocket motors, unlike liquid rocket engines, cannot be easily shut down after ignition, which is a major safety concern. Another drawback is that solid rocket motors are generally less powerful than liquid-propellant rocket engines, though once again this can be compensated for by simply having more propellant on hand in the form of a bigger motor.

Replacing the solid oxidizer of a solid rocket motor with a liquid or gas oxidizer produces a hybrid rocket engine. Hybrid rockets generally use a selection of easy-to-handle propellants, combining the simplicity of a solid rocket with the controllability and safety of liquid fuel engines. The winning X-Prize contender, Scaled Composites' Spaceship One, used a very simple hybrid rocket engine that went as far as dispensing with the overhead of being able to throttle thrust levels of the engine.[14] Hybrid rockets also combine the low performance of solid fuels with the problems of handling fluid oxidizers.

Seeking low-cost access to space through reusability is countered by the costs associated with refurbishment, as in the case of the U.S. space shuttle. Building one vehicle that can be used several times makes more economic sense than building several vehicles that can be used only once. However, from an operational perspective, the cost savings of not having to buy a launch vehicle for every launch must be balanced against the cost of reusing (or practically rebuilding) the launch vehicle. Returning a spacecraft to earth after it has achieved orbit requires that the kinetic energy involved with the orbit be dissipated in a safe manner. A huge amount of energy is required to get things into orbit, and this same amount of energy is involved in getting things down from orbit. The earth's atmosphere is readily available to convert a spacecraft's kinetic energy into heat, but this is a dangerous process requiring a combination of careful maneuvering and thermally protective materials. Features that allow for safe return—thermal protection, extra structure, flight control, and landing gear—all take up mass and bulk, which could be used for payload. In this sense, a reusable system is always less efficient when compared to an expendable launch vehicle built with the same technology. Once returned to the ground there is the process of preparing the launch vehicle for reuse. The technology of the space shuttle requires a ground crew that is often described as a "standing army."[15] If the technology does not allow for a cost-effective method of returning a launch vehicle to a flyable state, then there is no economic reason to design a reusable vehicle.

Launch vehicles have been designed since World War II. However, the design of a low-cost space launch capability remains elusive. The U.S. space shuttle is the product of design trade-offs made in light of the technology available at the time. For now, a low-cost launch vehicle has failed. The long and largely unsuccessful history of trying to replace the space shuttle only highlights the problems of technology not being able to live up to expectations. At the same time, without actually funding launch vehicle design, there is no way of knowing for certain if the technology is ready for cheap mass access to space.

Launch Vehicle Operation

All the major powers (and some medium ones, such as Brazil) have an existing or potential indigenous space launch capability. Indeed, any nation possessing a ballistic missile capability has the foundation for space launch, and many nations, like Canada, that do not possess ballistic missiles have the basic technological capacity to develop a space launch capability. Whether a specific nation decides to invest in such a capability, doing so successfully is another matter. The cost of space launch remains problematic, even though competition has resulted in some marginal reductions. Governments, and the large aerospace companies with government-subsidized research and equipment, have thus far been the dominant players in getting satellites into orbit.

Even though there are currently over 700 satellites on-orbit, space technology remains largely experimental in part because of the costs. Launch vehicles and satellites are far from being routine items even for military procurement. Spaceflight at present has more in common with launching an expedition into some far-off land than the sortie of a warplane. It is often claimed that by going into space more often the incentive to develop lower-cost space access (among other necessary cost-saving technologies) will finally break the present experimental model of spaceflight. This is the industrial model of space technology.

There have already been attempts to operate in space on industrial levels. Though financial failures, the huge satellite constellations launched under Iridium, and planned for Globestar and similar commercial ventures of the 1990s and early 2000s, show that the establishing of mass satellite constellations is not beyond reach. Prior to these commercial ventures, the U.S. Strategic Defense Initiative Office (SDIO) was contemplating the production and launch of hundreds of kinetic energy missile defense interceptors for the Brilliant Pebbles, and later scaled-down Global Protection Against Limited Strikes (GPALS), constellations. With the creation and maintenance of these satellite constellations, there was a perceived need for robust low-cost space access. Within both military and commercial worlds, several schemes to lower the preparation time, personnel needed, and total costs of operation were proposed. A few of these launch vehicle concepts continue to be promoted now after the collapse of satellite boom of the late 1990s.

Alternatives to space technology took away much of the demand for mass satellite constellations and attending launch capacity. Ground-based cellular technology has reduced the commercial viability of satellite phones, fiber optics have proved to be better links for most of the wired world than satellite broadband, and the Clinton and subsequent Bush administrations have redirected much of missile defense development toward ground-, sea-, and air-based systems. These are not the first examples of space projects being superseded by more down-to-earth equivalents, nor will they likely be the last.

In between the experimental and industrial models of spaceflight is a military sortie model. As spectacular as military aircraft seem to be in air shows, even the most expensive of operational high-performance aircraft can be said to operate in a routine manner. For most 21st-century major air forces, flying is expensive and dangerous, not from the act of flying, but instead from the missions being undertaken, which usually involve an opponent nation trying to inflict some harm on them.

A military model of spaceflight can be a way of justifying the expense of the current model of space launch. Given a severe enough interpretation of threat, national leaders will attempt to pay any price for security. The cold war resulted in what some consider absurdly large numbers of nuclear weapons, which under the concept of deterrence had the primary mission of existing and surviving to deter the other side from threatening core interests, and a secondary mission of being used in an all-out global war. Indeed, space launch in many respects is already operating in this mode; orbiting satellites started off as, and are still in many respects, strategic assets. During the cold war, military space programs did not stray far from the focus of the nuclear warfare mission. However, this military sortie model basically aims to have a space launch capability develop to the point where its use would be more akin to a special operations force. Special operations, or *silver bullet,* forces are not cheap, nor are they intended to be used haphazardly. Robust, routine, and reactive (on demand) spaceflight will not be cheap initially and therefore should also not be used haphazardly. The development of expensive but safe on-demand access to space is dependent on there being a convincing military need for this type of access. Otherwise, like the incredible pace of computer technology, the military may leave the maturing of space launch to the private sector.

Satellite Operations

The point of all the investment embodied by a launch vehicle is usually to get a satellite into orbit. A satellite can be divided into two parts: the mission payload and the bus. The mission payload performs the various benevolent and malevolent tasks done by satellites to produce the modern world of today and tomorrow. The satellite bus represents the rest of the satellite and provides the platform for which the mission payload can exploit the highest of high grounds. The function of the bus is to provide for the mission payload on-orbit maneuvering, power, thermal regulation, and command and control, which includes the link to the ground stations.

To counteract drifting caused by orbital perturbations and to stay in what its mission planners have defined as the satellite's useful mission orbit, the satellite must have some kind of on-orbit propulsion to perform active station keeping. In addition to station keeping, a satellite's onboard propulsion is also used to effect changes to a satellite's orbit. As most onboard propulsion is only capable of relatively low thrust levels, a satellite's ability to change orbit is modest, although with some creative course corrections, propellant, and time, major changes are possible. The limited amount of propellant carried by a satellite represents its life span. Once a satellite is out of propellant, it is effectively useless, regardless of the state of its mission payload. Moreover, satellite operators do not generally leave out-of-control satellites in prime orbital slots. A reserve of propulsion capability is kept to maneuver a satellite to reenter the earth's atmosphere in a safe manner, as was done with the Mir Space Station, or to push the satellite into a high disposal or graveyard orbit out of the way of useful orbits.

Another form of on-orbit maneuvering is control over the satellite or spacecraft's orientation (or attitude). While pointing a satellite in the right direction can be accomplished with rocket thrusters, various forms of gyroscopes can also be employed for attitude control. Gyroscopic stabilization and orientation control have the benefit of not requiring propellant, only electrical power. However these are mechanical devices, often experiencing continuous use, and will wear out over time.

Mission payloads are usually a collection of electronic devices and therefore must be supplied with power. Early satellites such as Sputnik 1 used batteries for their short missions. Longer-duration missions require the use of solar panels composed of arrays of photovoltaic cells and batteries. A satellite is not always in the sun. Therefore, enough rechargeable battery capacity must be hauled along to run the satellite in the dark. The solar arrays should have enough of a margin to both recharge the batteries and operate the satellite simultaneously. Solar power is effectively the only way to live off the land when in earth orbit and is the predominant power supply used by long-duration satellites.

For power levels beyond those that can be provided by the photovoltaic cells available at the time of a satellite's development, some have turned to nuclear power. The Soviet Upravlyaemyj Sputnik–Aktivnyj (US-A) series of radar ocean surveillance satellites (RORSAT) were nuclear powered. RORSAT nuclear cores were to be disposed of by propelling them to stable high orbits. However, this maneuver was not always successful, with one notable incident involving nuclear materials being spread across the Canadian Arctic in the late 1970s.

Future, more aggressive military payloads may also have large energy requirements. Chemical power, in the form of a laser beam generating reaction, is to be used in the proposed space-based laser. Other concepts have involved the use of explosive-driven pulse power generators and advance flywheel concepts. Notwithstanding certain interpretations of the 1967 Outer Space Treaty's clauses concerning weapons of mass destruction in orbit, harnessing the power of a nuclear weapon has been proposed in some extremely high-power applications such as the fabled bomb-pumped X-ray laser. These extremely potent power sources may be considered part of the payload as they are only used for the intermittent operation of a mission payload, leaving all other systems to be likely powered by the combination of solar power and batteries.

Thermal control for a satellite, the space-borne equivalent of air conditioning and heating, is absolutely critical to a satellite's operation and basic survival. Satellites generate waste heat, which must be carried away to prevent damage to the payload. Computer devices have a particular tendency to misbehave when they overheat, and most satellites are dependent on computers for basic operation. The problem in space is that the vacuum of space does not provide a medium for heat to be transmitted and carried away in. For the same reason that there is no sound in space, there is no such thing as wind chill in space. Excess heat can only be removed via the mechanism of thermal radiation where heat is transferred away from the satellite as infrared radiation. Having direct unfiltered sunlight shining on the satellite does not help matters, as the sun's radiant energy adds to the excess heat that a satellite generates. Sun shades and insulating materials help to cut down on external sources of heat.

On the flip side of the problem is the extreme cold a satellite can encounter in the shadows. Some batteries become erratic below a certain temperature, and some electronics fail to turn on properly if not warmed up. The space equivalent of a car block heater and blankets of insulating materials are required to prevent the satellite from freezing into a solid dead mass. Finally, there is the problem of knowing when to use heating and when to use cooling elements of thermal control. Early space research included study of the space environment—an investment that continues to pay off today for satellite design. Onboard monitoring and control over heating and cooling systems is often required despite the additional mass and bulk such systems add. Thermal regulation is a critical part of life support for a satellite's various electronic and mechanical components.

Control over a satellite is a combination of remote control via ground stations and, increasingly, onboard computers. While onboard computers are capable of monitoring and reporting on conditions, they are limited in their ability to make decisions. As such, ground controllers are still needed to oversee the health and operation of practically all satellites. The flow of information between the ground and the satellite represents an umbilical cord. In most cases, a satellite will not immediately come to grief if it is cut off from its ground control; however, over time most satellites will run into problems if housekeeping tasks are not ordered from the ground. More-sophisticated programming and more-capable on-board computers will over time reduce the amount of human supervision. However, it will not remove the need for human oversight. Most satellites represent too great of an investment to trust to pure autonomous computer control.

The satellite bus can represent the physical interface to which a mission payload is physically mounted. Some unique satellites demand that the satellite be built around the mission itself. This is costly as each satellite is a custom masterpiece. Major satellite manufacturers usually offer common bus designs within which certain bounds can be customized to a customer's needs. The majority of commercial communication satellites in geostationary orbit belong to one of the major satellite bus families. These still represent billion-dollar projects, with planned on-orbit life spans of over a decade.

Big satellites perform big missions, as for instance geostationary communication satellites (ComSat). Aboard each GEO ComSat is enough equipment to bridge a gap of approximately 35,800 kilometers to provide simultaneous service to multiple ground stations spread across continent-sized service areas. Large satellites are not mass-produced like cars and armored vehicles and certainly not churned out like missiles or for that matter even nuclear warheads. The procurement and total cost of large satellites put them in the same league as warships and the large merchant ships of international commerce, which at most are produced in small batches.

That is not to say that smaller satellites are not possible or desirable. In recent years, there has been a proliferation of small satellite buses, often prefixed with micro-, nano-, and even pico- to advertise their lack of bulk and, importantly, lack of cost. Due to the continued shrinking of computers, more and more capability can be placed into a smaller and smaller volume and mass. Smaller satellites have lower launch costs. This has led to a proliferation of developing nations and Western universities launching their first satellites. The lowered bar for space utilization means that these small satellites may be churned out like desktop computers or munitions.

Conceptually, satellite operations are often divided into segments. The satellite is the *space segment,* facilities for controlling the satellite constitute the *ground segment,* and the equipment needed to make use of a space system's broadcasts is the *user segment.* Sometimes space launch and satellite manufacture are listed as being segments as well. The exploitation of space now has a well-developed business model and organizational divisions. With potentially billions of dollars a year in revenue involved, the space industry has certain cause to develop its own set of best practices. Small satellites are representative of advances in technology shrinking the space segment, which may drive demand for smaller launch vehicles. The other segments are subject to the same opportunities. GPS user segments are now small enough to fit into wristwatches. Mission control, or ground segments, will always require staffing but now are housed in much smaller rooms than the great control rooms of the space-race era, as seen in the film *Apollo 13*. With miniaturization and deflation in costs, space capabilities are proliferating. For some, proliferation is the embodiment of the Outer Space Treaty's goal of space "for the benefit and in the interests of all countries." For others, the spread of space technology and know-how is a clear danger.

Conclusion

For everyone orbital space represents opportunity. However, not all opportunities are benign, nor are all opportunities wise. In the sometimes paranoid world of security and defense, every aspect of space utilization, from manufacture, launch, satellite operation, and ground stations to communication, links are all potential vulnerabilities. Access to space is proliferating, and this can only mean the security situation in orbit will become more complex. The same technology that will perhaps allow the rich to enjoy weightlessness in comfort, and later on the rest of us (probably crammed into economy service), will also give additional means of attack to those who wish harm to us and our way. At the same time, there are costs to seizing the *highest of the high ground*, which may be too high for any realistic return. Future opportunities of military space are a mix of science, technology, and policy. Technology and policy, given enough time and money to develop, can accomplish anything within the bounds of the physical universe. Money and time are products of the political and physical realities of nations.

Notes

1. National Aeronautics and Space Agency, "Reference Guide to the International Space Station," January 16, 2007, http://www.nasa.gov/mission_pages/station/news/ISS_Reference_Guide.html.
2. The Fédération Aéronautique Internationale notes for the U.S. Air Force SR-71 Blackbird, the absolute speed record of 3,529.56 kilometers/hour; Fédération Aéronautique Internationale, "General Aviation World Records," http://records.fai.org/general_aviation/absolute.asp.
3. Global Security, "SENIOR CROWN SR-71," http://www.globalsecurity.org/intell/systems/sr-71.htm/.
4. James Oberg, "Astronaut," World Book Online Reference Center, 2005, World Book Inc., http://www.worldbookonline.com/wb/Article?id=ar034800.
5. United States Air Force, "Factsheet: LGM-30 Minuteman III," November 2006, http://www.af.mil/factsheets/factsheet.asp?id=113.
6. Falling is described as gravity pulling two objects together, or the perception of the smaller mass being drawn down to the much, much greater mass.
7. Lucy Rogers, *It's Only Rocket Science* (New York: Springer, 2008), 303–306.
8. John E. Oberright, "Satellite, Artificial," World Book Online Reference Center, 2004, World Book Inc., http://www.worldbookonline.com/wb/Article?id=ar492220.
9. United States Air Force, "Fact Sheets: Defense Support Program Satellites," January 2008, http://www.af.mil/factsheets/factsheet.asp?fsID=96.
10. Daniel G. Dupont, "Nuclear Explosions in Orbit," *Scientific American* 290, no. 6 (July 2004): 100–107.
11. With scramjets, not only must the airframe be capable of hypersonic atmospheric flight, but it must also be part of the engine and still be lightweight enough to carry a worthwhile payload.

12. Larine Barr, 88th Air Base Wing Public Affairs, United States Air Force, "Pulsed Detonation Engine Flies into History," *Air Force Print News Today,* May 16, 2008, http://www.afmc.af.mil/news/story_print.asp?id=123098900.

13. National Aeronautics and Space Administration, "Candlestick Rocket Ship," January 23, 2003, http://science.nasa.gov/headlines/y2003/28jan_envirorocket.htm.

14. Scaled Composites, "Tier One Private Manned Space Program," http://www.scaled.com/projects/tierone/faq.htm.

15. Andrew J. Butrica, *Single Stage to Orbit* (Baltimore, MD: The Johns Hopkins University Press, 2003), 68.

CHAPTER 2

Military Space and Force Enhancement

Military space is an inseparable element of today's conventional warfare paradigm, regardless of the various concepts—the revolution in military affairs (RMA), the precision warfare, hyperwar, information war, the digital battlefield, or fourth-generation warfare, among others—used to describe modern war. Military space in its force enhancement role is arguably the foundation of America's military supremacy in the information age. Satellites have become the enabling technology for much of the precision, agility, and speed that has allowed the U.S. military to subdue those clinging to the old paradigm of industrial-age warfare. The significance of military space capabilities and U.S. space power in today's conventional warfare paradigm has its foundation in the classic theories of war by Clausewitz and Sun Tzu—the use of space to mitigate "fog and friction of war" and to fulfill the imperative to "know thy enemy and know thyself." Further underscoring the place of space in military matters has been the steady investment by allies, peer competitors, and those outright hostile to the United States to both emulate its space capabilities and counter its space power.

The Origins of Space Force Enhancement

In order to understand the significance of space in contemporary warfare, it is useful to examine briefly the evolution of warfare relative to technology over the past century. Warfare has always reflected the technology of the time. Industrial-age warfare followed from the Industrial Revolution of the 19th century in which industrial procedures and production were applied to warfare as evident in World War I. As industrial technology contributed significantly to the four-year stalemate and high level of casualties on the western front, so subsequent technological developments within the industrial mode of mass production (the assembly line) saw the return of mobility and maneuver in World War II. Blitzkrieg, or

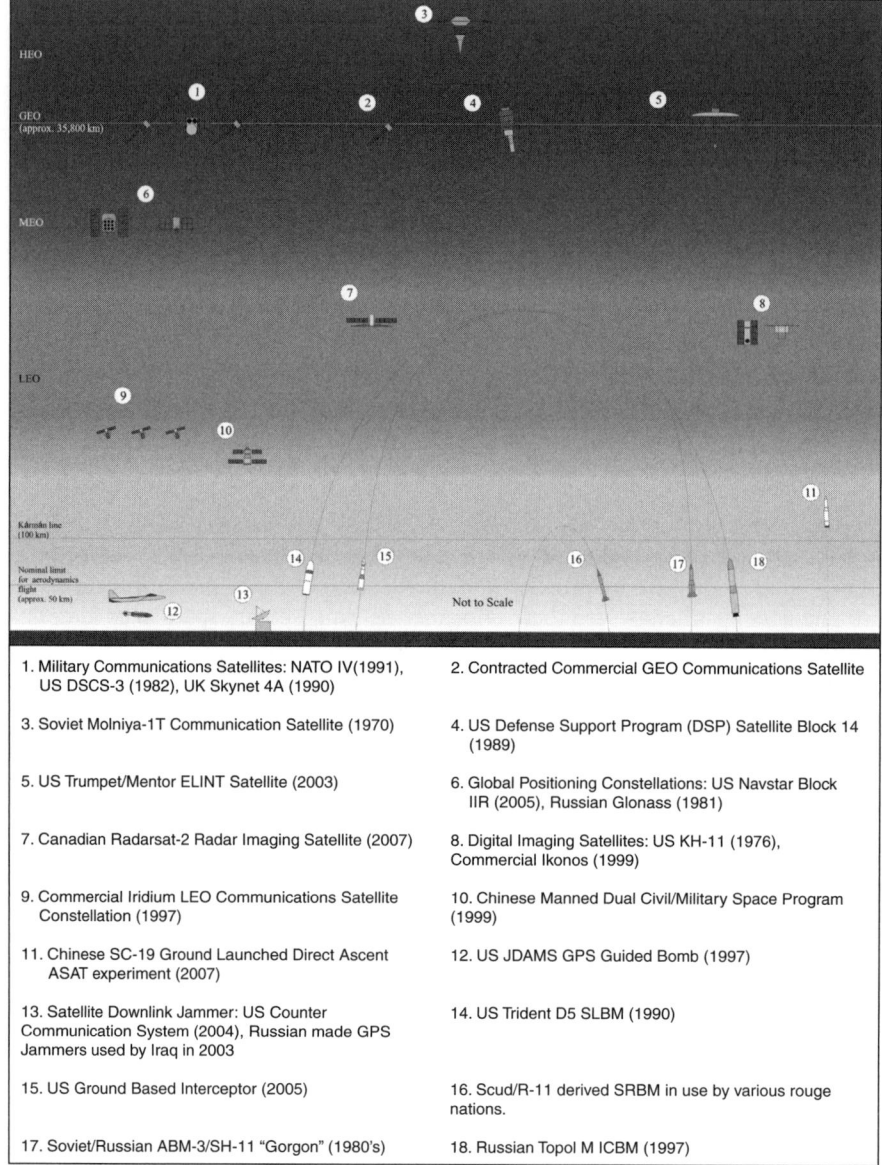

Figure 2.1 Present Military Space

lighting war, brought together armor, artillery, and aircraft via wireless radio to create a swift and powerful expression of combined arms. World War II also saw the beginning of the atomic or nuclear age where nuclear weapons became the primary focus of military matters for the following decades of the cold war. It was in the final decade of the cold war that emerging technologies, the microchip and computer

Military Space and Force Enhancement

network, put in place the foundation of the contemporary age of warfare, initially conceptualized as a revolution in military affairs (RMA). Though space technology has been around since the atomic age, with its own contending technological era known as the space age, it is in the information age that space has become an integral part of warfare.

The focused nature of information-age warfare contrasts with the mass destruction of nuclear warfare. Nuclear reactions release vastly greater amounts of energy than the most powerful of conventional chemical reactions. For this reason, the yield or explosive power of nuclear weapons is given in units of tons of the common chemical explosive trinitrotoluene (TNT). Due to the intricacies of nuclear physics, the first nuclear devices were in the kiloton (kt), or thousands of tons of TNT, range. The nuclear weapons that devastated Hiroshima and Nagasaki were 15 kilotons[1] and 22 kilotons,[2] respectively. With advanced nuclear weapon design techniques, such as mixing fission and fusion nuclear reactions in the same weapon, the yield of nuclear weapons was increased to ludicrous levels of destructive power, expressed in millions of tons of TNT—megatons. The realities of mass destruction drove strategy development in what was to be the centerpiece of the cold war superpower military relationship.

During the cold war, national security for the United States revolved around the many problems of facing another nuclear-armed superpower, the Soviet Union. The realities of the nuclear age led for the most part to the patterns of nuclear deterrence, and the well-known concept of mutually assured destruction (MAD). Essentially, in a MAD relationship in which both parties possessed the capacity to destroy the other even after a preemptive first strike by either one, the likely consequences of a large-scale nuclear exchange would have been an unacceptable amount of damage to all participants—mutual suicide. Thus, the MAD scenario produced a situation where there was severe disincentive to provoke a nuclear war. In effect, nuclear weapons produced nuclear deterrence as a war-prevention strategy.

Under the rational-actor model, one expects that a prerequisite for superpower status is some measure of rationality, so the consequence of the MAD scenario is arguably a stable nuclear standoff. The devastating consequences of nuclear warfare constrained the actions of the participants in the confrontation between the cold war superpowers. Even conventional warfare directly between nuclear-armed countries was a difficult proposition due to fears of it spiraling out of control, and these fears extended to regional and proxy wars beyond the central front in Europe. In the present day, even with reduced stockpiles of nuclear weapons and the emerging missile defense capability to defend against a limited ICBM strike, the potential devastation of nuclear weapons still tempers relations between nuclear-armed adversaries. Relatively recent nuclear powers, India and Pakistan, have cited the calming effect that their small nuclear arsenals have brought to their turbulent relations since independence.[3]

Providing a bridge between conventional warfare and the world of nuclear Armageddon was the tactical nuclear weapon—a nuclear weapon meant to be used against fielded forces, not strategic targets such as population and industrial centers or other strategic nuclear weapons. The rational for tactical nuclear weapons is simple—relative cost. Based on destruction per weapon, nuclear warheads are a relatively inexpensive means of waging war. Aside from an all-out nuclear exchange, the cold war had the potential to go hot in the form of mass-mechanized maneuver warfare across the central plains of Europe. The cliché scenario involved East Bloc armor and mechanized infantry pouring through the Fulda Gap, invading West Germany and beyond. The massive superiority in the number of tanks held by the Soviet Union lent credibility to this threat. To counter the Soviet superiority in numbers, the battlefield nuclear weapon option became desirable for quite some time. Though not always appreciated, nuclear weapons in some sense made up for the perceived gap in conventional forces needed to defend Europe from Soviet aggression. As a result, the United States, its allies, and NATO adopted a strategic policy of nuclear first use to bolster their deterrent postures.

As part of the Western view of deterrence (and not just against nuclear attack), the reason to be nuclear armed was to dissuade the Communist world from attacking the West. The resultant ambiguity over when nuclear weapons would be used was part of maintaining a credible deterrent. Deterrence would have failed if the other side had felt it could get away with a strictly conventional war. Years after the collapse of the Soviet Union, the rational for maintaining the nuclear deterrent and ambiguity over when it would be used, though diminished, remains, at least in latent form.

The history of the space age and the military use of space, as discussed in the Introduction, were intertwined with the atomic age and the cold war. Most of the successful space launchers of the era were converted nuclear delivery missiles. To this day versions of Sergei Korolev's R-7 missile continue to reliably launch payloads, including wealthy space tourists, into orbit. The space race itself was an outlet for cold war tensions—a place for each superpower to demonstrate technological prowess to reassure allies, win converts to the cause, and rattle the saber. In addition to demonstrating each superpower's skill to deliver payloads, the space environment was used in support of nuclear deterrence. The high vantage point of space was used for early warning systems meant to prevent surprise attack. Communication satellites contributed to the redundant links between command authorities and deployed nuclear forces. Orbiting reconnaissance satellites gave each superpower an idea of the other's capabilities, including the potency of the other's nuclear forces, which dispelled any misconceptions about the credibility of each other's nuclear capabilities. Though comparatively limited in numbers and capability, space systems were extensively used in support of nuclear deterrence, hence the superpower conflict that shaped practically all international and military matters of the cold war era.

While the bomb was the mainstream in military matters, the guided weapon was struggling to reach acceptance. The first guided weapons were deployed in World War II. Initially, their guidance was in the form of remote control either via wires or radio command. Among the World War II–era guided bombs was also a version of the German Hs-293, which experimented with a form of television guidance.[4] Like the atomic bomb, guided weapons concepts are meant to improve a platform's ability to destroy a specific target. Atomic weapons facilitate target destruction by providing a surplus of destructive force, increasing the acceptable margin of error when it comes to hitting the bull's-eye. Guided weapons on the other hand represent an economy of force—an attempt to apply force only where needed.

Technologies linked to the rise of the information age, solid-state electronics, the integrated circuit, and the microchip, allowed the evolution of guided weapons into truly useful and, more importantly, acceptable weapons. The electronics of the early cold war were bulky, expensive, unreliable, and consumed large amounts of power. All of these factors, especially the lack of reliability, made guided weapons at best a niche weapon, and for many a waste of resources better spent on proven systems such as nuclear weapons or enlarged conventional forces for an old-fashioned mechanized war. As electronic and related computer technology developed, equipment became smaller, cheaper, and robust and consumed less power.

Guided weapons, like nuclear weapons, offered the potential to reduce the number of attackers or sorties needed to destroy specific targets. With a guided weapon, or precision guided munitions (PGMs) as they are now labeled, city-sized regions did not have to be leveled to ensure the destruction of a single target. As the accuracy and precision of guided weapons increased, the amount of destruction that was needed to destroy a target declined. In the past, bombing missions were planned around the number of aircraft needed to destroy a target. Today, bombing missions are planned around the number of targets that can be brought under precision fire by a single aircraft.

The 1991 Gulf War, Operation Desert Storm, is often presented as the first demonstration of the effectiveness of PGMs. Though the majority of munitions employed in the 1991 conflict were actually of the unguided dumb-bomb variety, the general public was captivated by video footage of PGMs hitting specific targets. In contrast to the visions of megadeath invoked by nuclear warfare, the dropping of smart bombs came to acquire an antiseptic video-game quality. War, even with PGMs, is still a horrible and painful affair, but the promise of precision targeting to minimize destruction or collateral damage and to bring swift victory acquired great political appeal as a result of Desert Storm. Increased precision gave the option of limiting destruction to only that necessary to achieve victory, sparing both warfighters and bystanders from excessive risk and harm. The previously unimagined one-sided exchange rate in favor of coalition forces over those of Saddam Hussein, where the United States and allies suffered less than 300 combat-related deaths, certainly hinted at the promise of high-tech warfare to reduce the cost of war. Rightly or wrongly, the 1991 victory against Saddam Hussein was seen as a harbinger of how PGMs would change warfare in the future.

Along with the public's fascination with PGMs, the 1991 Gulf War is also presented by some commentators as the first space war. It was in this war that the public at large, and indeed many within the military establishment, became aware of how important the U.S. investment in the military space program was to overall U.S. military power. NAVSTAR Global Positioning System (GPS) satellites allowed ground forces to find their way through the featureless desert. The opening coalition air campaign was made possible by GPS, where GPS-equipped MH-53J

Pave Low (a large Special Operations transport) helicopters led U.S. Army AH-64 Apache attack helicopters, which at the time did not posses GPS,[5] to attack Iraqi radar sites in strikes carefully timed to open passages through Iraqi airspace for en route coalition aircrafts. The Defense Support Program (DSP) early warning satellites alerted coalition forces, as well as Israel and Saudi Arabia, of attack by Iraqi modified-Scud ballistic missiles. Though the Patriot Advance Capability–2 (PAC-2) tactical missile defense system, then under development, was unsuccessful in intercepting these missiles, the superior vantage point of space for operational early warning systems was demonstrated. Since the 1991 conflict, information technology has been informed by the lessons of this war and subsequent conflicts. The development of military space technology has been transferred from the nuclear warfare paradigm to that of war in the information age.

Alongside these dominant features of information-age warfare, and especially the key enabling role played by military space, is the increasing blurring of the age-old distinction between military technology and civil/commercial technology. In many cases, it is difficult to trace the specific relationship between the development and application of technology for military versus civil/commercial purposes. As noted in chapter 1, military and civil rocket technology developed in a hand-in-glove relationship. The origins of the Internet can be traced back to U.S. military developments to ensure continued communications in the event of nuclear war. Technology developed through investments by the U.S. Defense Advanced Research Product Agency (DARPA) for military purposes have regularly found their way into the civil/commercial sector, such as the computer mouse.

Military-developed technology applied in the civil/commercial sector is known as a *spin-off* relationship. In contrast, technology developed by the civil/commercial sector that is applied to the military sector is termed a *spin-on* relationship. There is also a long history of spin-on relationships, such as the creation of the tank from the tractor and the very origins of the airplane itself. While much of the initial technology related to space was driven by military investment, once NASA was created, civil investments and resulting technology would also be exploited for military purposes. In effect, space industries and organizations continue to be leaders in developing dual-use—civil/commercial and military—technology and application. The significance of digital computing with the rise of the microchip is a classic example of dual-use. The same factors that led to information-age warfare are also at play with the broader information-age economy.

In many ways, the commercial sphere has outpaced the military in creating and adopting information technology, in part because the battlefield is a harsher environment than the corporate boardroom, and the commercial marketplace is much larger than the military one. Regardless, the exploitation of advanced civil/commercial digital technology is central to the military use of space and has enabled military forces to use civil and commercial satellites for military purposes. Indeed, it is estimated that over 80 percent of satellite use by the U.S. military in Afghanistan and Iraq is derived from the nonmilitary sector.

Military Space and Information-Age Warfare

The information age has in some ways superseded the space age, with earthly technologies such as cellular phones, undersea fiber-optic cables, and personal computers having a greater public provenance in most people's lives than satellites, space stations, and missions to the moon. Fiber optics running directly to end users and short-range cellular coverage is perfectly fine for most domestic information applications. However, these relatively fixed technologies are too confining and insecure for a true global military power. Satellite constellations are the information age's means of power projection, using the commons of space to connect platforms on a global-scale battlefield. Space systems bring the effects of information technology to parts of the world where the information infrastructure is either sparse or nonexistent. The marriage of space with information technology in military affairs has led to the concept of space force enhancement.

It is difficult to trace the specific origins of the concept of force enhancement, although it is tied directly to the development of satellite capabilities of direct value to the conduct of military operations. It is the addition of a new military purpose for space systems alongside their original primary military function to support U.S. strategic deterrence. Generically, force enhancement simply refers to capabilities that generate or facilitate a more effective and efficient application of military force. Historically, one can conceptualize a range of military activities such as training, doctrine, equipment and logistics that play a force enhancement function in support of the warfighter. In providing this support, these activities also serve a force multiplier role, which has traditionally enabled smaller forces to defeat much larger forces as seen most clearly during the imperial age of European expansion into the Americas, Africa, and Asia. Indeed, the U.S. conventional military response during the cold war was to apply technology to offset Soviet numerical superiority in numbers during the cold war, even though the United States had the capacity to match the Soviet Union in terms of numbers. In this sense, advanced technologies served a force enhancement function, even though the term itself was not applied.

The first application of space systems to enhance military operations was the employment of weather data from earth observation satellites for planning operational air missions during the war in Vietnam. Since then, the marriage of new information technologies with satellites has resulted in a dramatic expansion of the role of space systems in support of the warfighter. While the basic force enhancement functions of space systems, dedicated military and civil/commercial satellites, are not new per se, their employment in space is, and it is this newness that spawned the term *force enhancement* itself. As a result, *force enhancement* when used today almost exclusively refers to the role of space systems. Even so, the idea of force enhancement also appears under different labels. For example, U.S. Space Command's 1997 Long Range Plan to 2020 does not employ the term *force enhancement* but instead addresses it in terms of the role of space systems in support of the range of U.S. military activities as defined by the Concept for Future Joint Operations—dominant maneuver, full-dimension protection, precision engagement, focused logistics, and information superiority.[6]

Space force enhancement is the collective term given to the use of orbiting satellites or systems to enhance the abilities of terrestrial platforms and munitions. The vantage point of space, reverently labeled the highest of the high ground, is the best location for the collection and rapid distribution of information. Precision warfare is dependent on having the best possible information available. This is information not only on threats and targets, but also on one's own fielded forces and those of allies. For information to be used effectively, there must be confidence in its accuracy, and it must reach those who can use it best. Space assets provide information support to earthbound warriors and weapons, increasing effectiveness not through increased force but by better management of the application of military force. While it is possible for terrestrial and airborne systems to be substituted for some space-based assets for regional coverage, only the vantage of orbit can provide the services needed for the digital battlefield on the global scale.

Space systems not only span continents, but also organizational hierarchies. Thanks to the information technology revolution, the direct impact of space systems has a much broader audience. The military use of space has steadily become ubiquitous, forming networks that link all levels from large-scale formations and deployments down to the individual soldier, sailor, and aviator. The reach of space systems in some cases extends to that of individual munitions and in future potentially to submunitions. Just as it is the vision for some information technology companies to network every element of a business or home together, it is the vision of many within advanced militaries to network as many elements of armed force together as possible—what is known as *net-centric warfare*. Space systems, as part of joint battlespace concepts, cross the boundaries of the separate U.S. armed services. Space systems allow disparate and mobile elements such as tanks, warships, and aircraft spread across a global battlefield not only to talk to one another, but also for one platform to engage a target only seen by another.

The revolutionary capabilities bestowed by space force enhancement capabilities raises the controversial issue of military use of space extending into the realm of weaponization, as discussed in detail in the next two chapters. For now, understanding the idea of space as a weaponized domain is dependent on one's perspective on what is a space weapon. In most respects, space is not commonly conceived as a weaponized environment. Yet, ballistic missile warheads have transited through space since World War II. Even so, the brief sojourn of long-range ballistic missile/warhead flight up to orbital altitudes has not generated a belief that military use has crossed into weaponization. The first generation of missile defense capabilities, the U.S. Safeguard and Soviet Galosh systems, which employed nuclear warheads at the upper reaches of the atmosphere, were not treated as space weapons. Even today, the newest generation of U.S. missile defense capabilities, which in February 2007 destroyed a deorbiting U.S. intelligence satellite that failed to reach its planned orbit, was not viewed in space-weaponization terms.

Nor are the traits of space force enhancement and other information-age warfare concepts linked to weaponization. Nonetheless, the widespread and very open use of space systems to aid in putting steel on target has made the distinction between space weaponization and the rest of the military use of space somewhat fuzzy to say the least. Quite clearly, the question of space weaponization is not just a far-off problem hidden behind technology that exists only in the minds of scientists and science-fiction writers.

Earth Observation Sciences

Earth observation sciences are often the unsung heroes of a nation's defense. Superior knowledge of a battlefield's characteristics is often the deciding factor in victory. The characteristics of a region, its terrain and weather patterns, drive what options are available for both politicians and military commanders. Understanding the options available goes a long way in getting into the other side's head. It is always preferable to exercise the option of choosing one's battlefield and gaining tactical home-field advantage. For example, the Duke of Wellington's survey of the terrain around the village of Waterloo, and his resulting choice to fight at Waterloo, significantly contributed to Napoleon's ultimate defeat in 1815.

Since the beginning of map making, commanders and planners have been dependent upon maps for their campaigns. The flaws of maps are also telling in indicating lapses in knowledge and critical pieces of topographical information that could decide victory or defeat. Knowing the geography of an area guides the planning of a campaign, and discovering in the midst of battle a previously unknown geography often defeats planning. Knowledge of the terrain allows for speed, safety, and efficient movement. Thus, it is no surprise that many of today's survey organizations and observatories have military roots. For example, the United Kingdom's chief mapping organization, the Ordinance Survey, has a history dating back to the Napoleonic Wars and concerns over how to defend the island nation from French invasion.

Whereas in the past map making was a terrestrially restricted activity, space provides a vantage point to survey the entire earth—a necessary capability if one is projecting military power and defending and attacking on a global scale. Satellite imagery of the earth has greatly simplified the task of map making. Basic pictures of the earth, combined with elevation data from satellite-based sensors, produce on demand high-fidelity representations of the earth's physical surface. One application for this type of detailed topological data is autonomous navigation systems, such as the U.S. terrain contour matching (TERCOM) system found on cruise missiles prior to the establishment of ubiquitous satellite navigation. In the 1991 Gulf War, cruise missiles had to take the scenic route through mountainous terrain along the Iran-Iraq border to obtain geographic fixes not available in the comparatively featureless deserts found in the vicinity of most Iraqi targets. While this need to detour over land with missile visible landmarks represents a definite limitation to indirect satellite support, TERCOM and similar systems are completely autonomous and immune to jamming or loss of navigation satellites.

A more current example of the application of satellite-generated maps is the 2001 campaign to remove the Taliban regime from power in Afghanistan. Under the Taliban, Afghanistan was a closed-off country with its people kept away from the modern world and modern ideas. Unlike the West, there was no domestic demand for up-to-date road maps that could be exploited by the United States. Traditional land survey and cartographic resources for Afghanistan were limited, and the United States faced a lack of knowledge of the terrain. As a result, the United States turned to commercial satellite imagery, which had achieved some measure of commercial success in the years prior to 2001. As a function of digital transmission technology, the United States could obtain images of specific target areas constrained only by the nature of a particular imagery satellite's orbit. Moreover, depending upon orbital paths, maps could be updated periodically during the conflict. Seeking to ensure operational secrecy over the Afghanistan theater of war, the United States bought exclusive rights to all commercial imagery of the region—a step it had also taken during the Gulf War. As a result of these exclusive contracts, the United States and its allies were more than well supplied with fresh maps of Afghanistan prior to the commencement of hostilities.

The importance of satellite imagery for obtaining accurate topographical information for campaign planning and execution is a double-edged sword. Commercial imagery is available to anyone willing to pay, unless the company producing the imagery is explicitly prohibited by national law from selling such imagery to certain clients or nations. If the U.S. government had not obtained an exclusive contract to commercial earth imaging data of Afghanistan, its adversary potentially would be able to acquire such data, undermining U.S. information superiority. Commercial satellite imagery may have fallen into the hands of the Taliban and its supporters. Despite promoting a very austere form of Islam, the al-Qaeda organization, which had training camps in Afghanistan hosted under the Taliban regime, has no qualms with using technology to achieve their aims, including the possible use of commercial space technology. In addition, the media could also have potentially accessed the imagery to report on the war, thereby affecting U.S. operational secrecy. Indeed, in early 2001, commercial space imaging had been used by the media to obtain pictures of a U.S. Navy EP-3 in the middle of a Chinese airbase after a midair collision with a Chinese fighter forced it to make an emergency landing in China. U.S. contractual control of overhead images from commercial satellites also enabled the United States to provide maps of the war zone to the media without disclosing the capabilities of U.S. satellite image reconnaissance capabilities and ensure that the maps did not disclose any sensitive military information.

With computer technology there is also no reason to stop with just emulating old mapping products. To assist in visualizing the prospective battlefield, elaborate models are often constructed, complete with scale representations of the opponent's defenses. Today, through the use of computers, accurate satellite-derived topological data can be used to build virtual representations of the battlespace. These three-dimensional representations can be inserted into various military simulators providing the opportunity for low-cost, high-fidelity mission training without risking personnel or equipment. This technology has also made its way into civilian life by filtering into hobbyist flight simulations and other entertainment software.

Further information can be embedded into digital maps, including political, economic, and environmental data. Satellite-aided digital maps and tools for manipulating data are now nearly ubiquitous. Detailed maps of most of the earth's surface are not just available to government users (military and civil servants concerned with a particular piece of ground) but are now free for the taking by practically everyone with access to the World Wide Web. The Internet company Google has not only made satellite imagery available online, but has also integrated it with various mapping tools ranging from simple addresses to automated route planning and, most recently, combining the overhead imagery with links to street-level imagery. Data and integrated utilities to make use of the data are continually increasing, in part as a function of competition among Internet companies. GoogleMaps and similar free Internet mapping services, such as MapQuest, have become convenient means for individuals to locate landmarks for navigating around the city they live in. These developments, in turn, reinforce the utility of commercial products for military use.

Alongside maps, space-based imagery is vital for weather reporting. Meteorology, like cartography, has been central in human conflict. Weather, like geology, can cause grief to navigation and military enterprises, and the weather is much more prone to rapid change than geology. Weather conditions control what operations can or cannot be conducted. During the Battle of Britain, the Royal Air Force only had to sustain control of the skies over southern England long enough for seasonal weather changes to make air attacks difficult and an invasion across the English Channel impractical. On the eve of the D-Day invasion of France in June 1944, Allied planners agonized over weather conditions over the English Channel. On a smaller timescale, immediate weather conditions affect day-to-day air operations. Poor weather may delay or outright cancel operations. At sea, blundering into severe weather may be as destructive as any naval weapon short of a nuclear warhead.

Meteorological science seeks to mitigate the effects of weather by providing both long- and short-term predictions of environmental conditions. Earth orbit provides an excellent vantage point for monitoring the development and movement of weather on a regional and local scale, as most publicly evident in the release of satellite images of hurricanes as they develop and move. Weather satellites are built around various sensors meant to see weather, including optics across a wider spectrum than is visible to the human eye. Superior information on the weather, to the point of near real-time monitoring in some cases, allows for superior planning to mitigate the effects of weather.

Of importance to precision warfare, weather conditions can disrupt the operation of optically guided and laser-guided smart weapons. Despite the proliferation of GPS and millimeter wave radar guidance, laser- and optically guided weapons still have a place in a technologically sophisticated arsenal. For instance, recent improvements to the joint direct attack munitions (JDAM), commonly thought of as a GPS-guided weapon, include the addition of a laser guidance seeker to allow precision engagement of moving targets. The JDAM's laser guidance option is meant for use in clear weather. The original GPS and inertial guidance are available for use in all weather conditions where bombs may be dropped.

Space-based earth monitoring also extends beyond the surface of the earth to weather in space. For example, it is believed that ICBM guidance systems utilize data on local variations in the earth's gravity (due to imperfections of the earth from a perfect circular sphere) to enhance accuracy. Gravity and other geodesy measurements over the flight path of an ICBM, including the parts over the target country, can only be safely obtained from orbit. As a result of improved accuracy derived from satellite data, some have argued that ICBMs and submarine-launched ballistic missiles (SLBMs) could be employed as conventional, precision strike weapons. Of course, this is highly controversial because of the problems differentiating between an ICBM carrying a conventional warhead and one carrying a nuclear warhead in a situation where an adversary also possessed nuclear weapons and feared a possible disarming first strike. Indeed, such precision as enabled by space assets, whether from ICBMs, cruise missiles, or air-launched JDAMs, for example, all hold the possibility of a disarming first strike without using nuclear weapons. In effect, space systems' applications, as further discussed below, have been central to the replacement of the deterrence paradigm by the precision conventional warfare paradigm. The full strategic and political implications of this dramatic change have yet to be fully addressed and will likely be a major focus of future strategic assessment.

Surveillance, Reconnaissance, and Targeting

Military terminology distinguishes between three types of observation: the surveillance of a wide area of territory or military activity, the reconnaissance of a specific area of territory or military activity, and the targeting of specific territory or military activity for attack. In one sense, these three correlate to the concepts of strategic (surveillance), operational (reconnaissance), and tactical (targeting). In another sense, they are interrelated. Surveillance of a wide area identifies activity of operational interests on the part of the adversary and cues reconnaissance of particular threatening operational activities, which, in turn, identifies potential targets for the direction of ordinance. All three have been greatly facilitated by the highest of the high ground—space. Moreover, the continued refinement and proliferation of orbiting earth imaging capabilities has blurred somewhat the line between strategic surveillance, reconnaissance, and targeting intelligence gathering and tactical observation. In some specific niche roles, space-based sensors now have the potential to participate directly in the sensor to shoot cycle of a weapon engaging a terrestrial target. While the highest of the high ground is perhaps intuitively the best vantage point for military observers, the laws of nature governing orbiting have limited just how useful a satellite can be in engaging targets on earth. The march of technology is, however, hinting at ways to overcome nature for satellites to participate in the sensor to shoot cycle.

Objectively there is not much difference between an earth imaging satellite and an image reconnaissance satellite. Like many dual-use technologies, the separation between tool and weapon is a matter of intent. Commercial earth imaging has already proved itself for map-making applications. The fear that commercial earth imaging services may be employed by those hostile to the United States clearly underscores the reality that even limited civilian capabilities are of potential intelligence value. Even so, the capabilities of National Reconnaissance Office (NRO), the agency that runs the majority of the U.S. orbiting reconnaissance programs, are estimated to be of higher resolution than those offered on the open market.

The development of orbiting earth imaging capability has in many respects mirrored the development of consumer camera technology. Both the consumer-grade camera and the spy satellite camera are based on the same laws of physics governing optics, except the orbiting camera is on a much larger scale than the camera found on any vacationing tourist. Initial U.S. imaging reconnaissance satellites utilized photographic film cameras, just as the cameras of the era. The first of these satellites run under the CORONA program was launched in 1959. These satellites used the now fabled KeyHole (KH) designation to separate series of imaging satellite types. As film was used to record images, these satellites had to employ small reentry vehicles to return pictures to the earth, which were captured by a plane as it descended by parachute back to earth. While there is no commercial equivalent to the early film return imaging reconnaissance satellites, obsolete images from successive series of KH satellites have been released to the public domain. Starting in 1995, select images from the KH-1 series of satellites were made available. The most recent declassification of U.S. satellite imagery occurred in 2002 with the release of imagery from the KH-9 HEXAGON, or more commonly Big Bird, series of satellites.

The nature of the recording medium, film, restricted the operation of these types of satellites. As pictures taken by early KH satellites could only be obtained from the satellite with the return of the film return reentry vehicle, operators had to decide on using up the entire roll of film before commanding the film capsule to return or wasting unused film to return some critical pictures early. Film retrieval also exposed the valuable pictures to all the hazards of reentry. Finally, the quality of the images or if images were actually taken would not be known until the film was developed. Just like the roll of film inside the ordinary tourist's camera, each roll of film on the spy satellite only had one chance to produce pictures worth keeping. The limitations of film were partially mitigated by equipping KH satellites with multiple film return capsules. Starting on the later KH-4 CORONA satellites two film return capsules were carried. The final film return spy satellites, the KH-9 Big Birds had five.

Electro-optical sensors and digital computers made digital photography available first to satellites and much later to the consumer. While there remains some truth to claims that analogue film can produce a superior picture, both in the civilian world and in the national security world, digital-imaging technology does have the advantages of speed and flexibility. Instead of recording images on film, image data in digital form is readily transmitted back to ground stations. This meant that the KH-11 series of satellites, which used an electro-optical imaging system, could be designed for a much longer service life in orbit due to no longer being limited by how many pictures it could take before running out of film. Digital transmission of pictures also meant that the response time for obtaining pictures from a spy satellite was only limited by the nature of a satellite's orbit relative to the location of ground stations and whether the controllers thought it was worthwhile to use valuable propellant to adjust the orbit. Digital imaging is the enabling technology for the Hubble space telescope, which some have claimed bears a physical resemblance to KH-11.

The same orbital digital-imaging technology in conjunction with the explosive proliferation of computer technology in home and business has allowed for the viability of the commercial earth imaging as a business, although with a much lower resolution. Recent commercial imaging satellites are, however, closing the gap with their national security counterparts. In reverse direction, the militarization of cutting-edge civilian information technology opens up the possibility for the increased distribution of imagery intelligence. Once a picture is in digital form, it can be copied and distributed as needed.

While the turnaround time of today's imagery intelligence satellites is much faster than that of film, and the distribution of imagery intelligence increased, its abilities to affect the tactical battlefield are still limited by the science of orbiting and photography. The quality of pictures from a camera operating in visible wavelengths, whether in the hands of a tourist or in space, is limited by lighting. A vacationer has the option of having a flash to illuminate the subjects of low-light holiday snaps. This is not an option available to the spy satellite. A satellite-based optical camera can only take pictures when the sun is able to provide lighting and there is no cloud cover obscuring the area or target of interest. Thermal-imaging cameras allow for low-light and no-light imaging capability but still cannot penetrate some cloud cover. Synthetic aperture radar (SAR), the use of a moving radar and computer processing, allows radar to produce images of relatively high resolution and is not affected by cloud cover or lighting conditions. Canada's RadarSat I and II are prominent examples of SAR-based satellite surveillance imaging.

Ultimately, the number of operational imaging reconnaissance satellites in current use has resulted in their employment as strategic assets, instead of assets for direct targeting purposes. With the increased life span of imaging satellites for strategic purposes, not many are needed. KH-11 satellites and their near-term successors have life spans measured in years; they cost billions, and there are only a few operating in orbit. The orbits of these satellites are relatively low and therefore have a field of view that covers only a small portion of the earth's surface. If they were higher up so that the field of view covers a hemisphere, then the resolution would be so poor as to not be able to see anything of strategic significance. This, of course, can be addressed by more powerful, larger optics, but space launch capability and costs remain a severe constraint on how big a reconnaissance satellite can be lofted. Moreover, the small number of satellites means that tasking these reconnaissance assets is controlled at the highest levels. Although a component of the information revolution is to give access to lower and lower levels of an organization's hierarchy (flattening it in the process), the operational model in use today for spy satellites simply cannot fulfill this empowerment role. These imaging reconnaissance satellites are of course meant to collect strategic-level intelligence—to study ships under construction in shipyards, to catch a glimpse of a new aircraft caught out in the open, to provide before-and-after images related to deployments and movement, or, in an almost tactical, use to count parked tanks in a marshaling area.

For tactical optical intelligence, the atmospheric unmanned aerial vehicle (UAV) has seen much greater proliferation than satellites. Larger atmospheric systems are relatively cheap enough to be somewhat plentiful, which results in control being released to lower levels of the chain of command. General Atomic's Predator-B UAV, for example, embodies the whole sensor to shoot cycle by carrying its own weapons to engage directly any targets it may find as determined by its controller. Small man-portable UAVs allow infantry patrols to carry their own organic aerial surveillance, relaying imagery back to small portable computer devices. This class of micro-UAVs is cheap enough that it may be procured in such quantities to give everyone who needs one their own aerial spy. In this respect, the potential for widespread space-age satellite-derived tactical image intelligence is upstaged by information-age computer technology and robotics. Of course, UAVs are also vulnerable to a range of air defense capabilities.

Despite the proliferation of UAVs, today's reconnaissance satellites do possess operational value, and the strategic, operational, and tactical realms are converging. Not all tactical targets are small and hard to see from orbit. If a target can get the attention of a satellite in orbit, then the satellite may be used for tactical applications. The Soviet Union during the cold war launched several nuclear-powered radar satellites, RORSAT, for tracking shipping and warships, which were used, for example, to find U.S. aircraft carriers, among other purposes. In the event of war, located carriers could be attacked by long-range cruise missiles launched by aircraft, surface ships, and submarines. The Soviet (now Russian) Project 949 submarine, or "Oscar Class" in NATO coding, is noted as having satellite detection as one means to give initial targeting for its powerful antiship cruise missiles.[7] Space, through the RORSAT, was to be used by the Soviet Union as part of its sea denial strategy to counter U.S. sea power.

Early warning satellites, which are a form of reconnaissance satellite, are designed to stare at the earth for the easy-to-see signatures of a rocket in powered flight. The U.S. DSP satellites arranged in a constellation in high geosynchronous orbits provide continuous coverage of a wide band of the earth's surface. Even at these high attitudes, the bright heat signature of a long-range missile with its engines still burning is detectable. DSP was originally a strategic asset to provide early warning of Soviet ICBM attack. Since its employment in the Gulf War, it has also acquired an operational, theater role and, through the evolution of its role to cue missile defense systems, a tactical role as well.

DSP is able to detect the hot engine plume of a ballistic missile from geosynchronous orbit, allowing for a small number of satellites to provide near global coverage. It is also rumored that DSP, and its replacement, the space-based infrared-high satellites, are also able to detect hot engine plumes of a lower intensity than ballistic missiles, such as aircraft, if calibrated to do so. Regardless, less-intense target signatures require the use of more-powerful sensing technology or lower orbits. Lower orbits require more satellites to provide the same coverage. Small satellite technology is being touted as being a possible platform for providing space-based tactical surveillance.

One of the most successful tactical space force enhancement applications has been the U.S. GPS. As an omnipresent global system, the space-borne architecture for GPS is based around a relatively large constellation of 24 active NAVSTAR satellites in medium earth orbit (MEO). The GPS system, created and maintained at great cost to the United States, provides accurate navigational, positional, and timing data to anyone with a GPS receiver at no cost. Today, GPS is critical to the shipping and transport industry. Combined with digital mapping technology, GPS gives waypoint-by-waypoint directions to delivery services, taxis, emergency services, and ordinary drivers. The GPS timing signal is also used to synchronize financial transactions down to the level of the corner bank automated teller machine (ATM). The low cost of GPS receivers has also meant that new civil applications are being found on a continuous basis. For the military, the originators and custodians of the GPS system, it has become a means to victory.

GPS is based on three components: a space segment, a control segment, and a user segment. The complete space segment is the constellation of 24 NAVSTAR satellites arranged into six orbital planes in circular MEO. GPS satellites in addition to their navigation and timing payload also carry detection equipment meant to catch sight of and pinpoint above-ground nuclear explosions. The control segment for GPS, run by the U.S. military, is made up of ground stations that monitor and update GPS satellites to ensure the accuracy of timing and positioning data. Finally, the user segment is anyone or anything with a GPS receiver, ranging from a ship at sea, an aircraft, or an orbiting spacecraft to a GPS-equipped cellular phone. Each satellite continuously broadcasts position and timing information in two forms or codes—a standard positioning service (SPS) level for civilian use and the classified precise positioning service (PPS), or P-code, primarily for military use. Line-of-sight to four satellites is needed to calculate position in the three physical dimensions along with time being the fourth. The more satellites visible to a receiver, the greater is the accuracy. GPS signals, however, can be selectively degraded, which limits access in specific regions. While the United States has incentives to degrade the civilian signal in areas of military operations where adversaries might employ GPS for their military operations, the Clinton administration pledged no to do so, regardless of the situation, in part because too much economic activity depends on the availability of this service.

The accurate positional and timing information supplied by GPS is an enabler for forces on the battlefield. GPS allows unprecedented levels of coordination on the battlefield in the maneuvering of forces right down to the lowest level. Friendly, or blue force tracking (BFT) and similar systems automatically collect and report the position of the vehicle or even the soldier to which it is attached. From a composite of the positional data, a commander can have an accurate picture of friendly (equipped) forces and how they relate to each other on the battlefield. The confidence GPS gives on the location of one's own forces reduces the fog and friction inherent in the operational art of wielding a fighting force.

The basic concept of blanketing the earth in navigational information has also simplified many of the problems of autonomous flight and allowed for explosive growth in the number of PGMs. Autonomous flight, that is flight without direct human input, is a capability possessed by many UAVs, particularly in longer-range platforms. Instead of some horribly complex and expensive navigation system, a low-cost GPS receiver simply tells the robotic air vehicle where it is in time and three-dimensional space. With such data, the UAV's onboard computer logic can easily make decisions on what it has to do to get to a specific place on time. These advanced UAVs only have to be supplied with coordinates and instructions for what to do when they are achieved and the UAV can carry out its entire mission without human intervention. On August 24, 2001, a U.S. Global Hawk bearing the name *Southern Cross II*[8] landed in Australia, successfully completing a milestone autonomous flight from Edwards Air Force Base.[9]

An aerial guided weapon is essentially a UAV with a warhead. GPS guidance in cruise missiles is a lower-cost alternative to terrain matching systems, and inertial navigation systems (INS), which lose accuracy over time. The joint direct attack munition (JDAM) uses a combination of GPS and INS to achieve all-weather precision guidance. Low-cost JDAM tail kits can be attached to existing stocks of iron bombs, allowing for conversion into smart bombs. The JDAM system's circular error probability (CEP) or accuracy is given as being 5 meters when GPS is available.[10] This means that at least half the JDAMs will fall within a 5 meter radius (10 m diameter) circle around the aim point, as long as GPS is working. Without GPS, the JDAM can still function with some accuracy as INS guidance can function without external updates from GPS. However, the accuracy of INS is dependent on the quality of the position information from the point the bomb was dropped, and the time it takes the JDAM to reach its target, as INS looses accuracy with time from its last external fix. The CEP without GPS for JDAM is 30 meters with a maximum flight time of 100 seconds from release after its last external location fix.[11] The CEP difference between GPS and INS is significant, especially in relation to the amount of explosive charge needed to destroy a target and extent of collateral damage.

The political push to reduce collateral damage demands that smaller amounts of explosive are used, which in turn increases the need for accurate and precise weapons delivery. The U.S. Small Diameter Bomb (SDB) program aims to produce a 250-pound weapon guided by a combination of GPS, INS, data link, and eventually laser designator terminal guidance. This is half the weight of the smallest JDAM—the 500-pound GBU-38/B. The small size of the SDB family of weapons allows for more bombs to be carried per sortie or platform. However, less-explosive power means a smaller margin of error to achieve equivalent effects of bigger bombs. Only GPS can provide all-weather guidance with the precision needed for SDB and similar weapons to be effective.

The war against the Taliban in Afghanistan led to a somewhat curious assignment of strategic bombers, like the U.S. B-52, in a tactical close air support (CAS) role. As background, in the U.S. Air Force there has historically been internal rivalries and outright conflict between the various mission communities. Strategic bomber advocates vied with the tactical air communities for funding and influence. Within this environment, the CAS mission had to fight for survival among the fast-jet advocates and the strategic air power types. With the technology available in 2001, JDAMs and portable GPS target designators, combined with the unique situation of the Taliban having practically no air defenses, allowed all elements of offensive airpower to participate in the critical precision support of U.S. and allied forces on the ground. Strategic heavy bombers flew high overhead waiting for the call from the ground to drop GPS JDAMs on targets specified by forces on the ground. Instead of carpet-bombing an area, these precision strikes were called in by forces directly in contact with the enemy. The large bomb loads of aircraft such as cold war–built B-52s allowed these aircraft to circle overhead for hours providing CAS. Space force enhancement gave a new and important purpose for legacy systems in the respect.

Competing GPS constellations are also emerging. Despite the United States providing the world access to its own GPS as a global public good, and having committed to allowing generally reliable civilian access, the nature of the international system means that those able to set up their own GPS equivalent have a tendency to do so. The Russian GLONASS GPS program was inherited from the Soviet Union. The GLONASS constellation has been largely reconstituted after over a decade of neglect due to the lack of funding for replacement satellites. It should shortly, if not already, be able to provide full coverage. That being said, the U.S. NAVSTAR GPS constellation was incomplete during 1991's Desert Storm but is still recognized as one of the contributing factors to the U.S.-led coalition's overwhelming victory over Saddam Hussein. The European Union's soon to be launched Galileo system is another source of irritation and outright alarm for the U.S. defense establishment since it offers NAVSTAR GPS-like capability outside of their control. Chinese investment in Galileo only adds to fears that it will someday be used against U.S. interests. China, for the same reasons that Europe and Russia want their own independent GPS system, is also developing a regional navigation satellite system independent of its participation in Galileo. There are ongoing discussions and working groups between the United States, Russia, and the European Union to coordinate so that the systems' services do not inadvertently degrade services to users.

The Movement of Information

In military operations there are two realities with regard to information—the distribution of information to those who can make use of it and the nearly insatiable appetite for more and more information. Speed, agility, and precision in military operations are products of having not only the right information to make battlefield

Military Space and Force Enhancement 59

decisions, but also the rapid delivery of time-sensitive information to the decision makers. It is in meeting these realities that communication satellites play a vital role in U.S. space force enhancement. For mobile forces deployed in far-off battlefields, only communication satellites (ComSats) can provide the data links critical to information-age warfare.

In the bent-pipe communications model, the satellite is a relay point between two points on earth—the satellite being the elbow of the metaphorical communication pipeline. The bent-pipe model allows for communication beyond the horizon. Due to the spherical shape of the earth, the maximum distance that the bent-pipe model can span conceptually is limited to a hemisphere. Real-world considerations actually reduce the coverage to less than a hemisphere. Three satellites are needed to always maintain at least one above the horizon in a position so that a signal can be relayed off it. Operationally, the higher the ComSat's altitude, the greater the coverage it can provide. Geostationary orbits (GEO) give a ComSat both the necessary altitude and the added characteristic of having its position relative to a point on the equator fixed. The distance between the earth and a GEO ComSat, 35,800 kilometers at its closest point, however, means that more power is required by both the ground stations to get a legible signal up to the satellite and the satellite to rebroadcast the signal down to the destination ground station.

Sacrificing the benefits of having the ComSat in a fixed position in the sky, orbits below GEO are used to reduce transmission power requirements at the cost of reducing satellite coverage. To maintain continuous coverage, a much larger satellite constellation must be employed. From the minimum of three satellites in the GEO ComSat, a LEO satellite constellation requires dozens and in some proposals hundreds of satellites. In the civilian world, the promise of true global coverage has been outweighed by lower costs of terrestrial communication technologies, resulting in the collapse of Iridium, the first LEO ComSat industry, before it even really began. The Iridium service managed to get its constellation of 66 active satellites, plus some spares, into orbit before financial difficulties curtailed the industry as a whole. The U.S. military stepped into the financial breach, and is now a major user of the Iridium service, giving fielded forces cellular-like portability for voice, messaging, and data.

In a curious twist of technology, the long-range UAVs, which have denied satellites a major place in the near-term future of tactical reconnaissance, are themselves dependent on satellites for communications. A common characteristic of the Predator, Global Hawk, and other long-range UAVs is a hump located on the top, often emulating the shape and position of the bubble canopy found on fighter aircraft. This is the aerodynamic fairing for a satellite communication antenna. While developments in robotics were increasing the potential for autonomy, the information revolution was binding things closer together into real-time networks. Unlike pre–digital age spy planes, UAVs are capable of supplying data through satellite links to end users in near real time. Just as in the civilian world, the clients of aerial intelligence gathering have become accustomed to faster and faster service and simply cannot wait for film to be developed and copies made.

The more information a UAV is capable of gathering, the more bandwidth is needed, and hence a relatively large antenna is fitted, necessitating the prominent cockpit-like hump. As the data link between the UAV and its controllers goes two ways, these aerial spies can be rapidly retasked, giving these platforms unprecedented flexibility. Conceptually, the UAV's data link can be routed to fielded units, or to a National Command Authority, making it part of the network of systems that allows a nation to project information superiority. Without satellite communications, today's attention grabbing UAVs would be limited as an information-age platform.

Modern society is in large part defined by connectivity—the degree to which information technology (IT) binds and relates everything and everyone together. Since the 1990s, the Internet has continually connected more and more people together—a fact that has had significant societal impact. Information is power, and individuals having greater access to information means that individuals have the potential to be more powerful. In the other direction, IT allows large organizations to act with precision on the small scale. As Thomas L. Friedman puts it, "the small shall act big . . ."[12] and "the big shall act small,"[13] rules that are also applicable to the defense world. Long-range UAVs such as Predator or Global Hawk are examples of this information-age phenomenon. The imagery supplied by a Predator UAV can be simultaneously displayed to a commander or even an individual soldier in the middle of the battlefield and the commander in chief back home. It goes beyond simply sharing reconnaissance data, as IT tools allow and encourage collaboration. UAV imagery can be combined with other data to provide context to what is being viewed. The tools needed to tag collected information with user-specific notes and to share such analysis are available along the entire length of the command chain. Indeed, it can be used as a feedback mechanism via satellite links to allow those on the front to relay back information to those leading from the rear.

The corporate world has seen the benefits of global-scale collaboration on productivity, agility, and speed. Instead of working independently, business units are expected to connect together. Instead of individual computers working in solitude, they now work in networks. However, the business world rarely ventures away from the infrastructure of the globalized world. Influential former Pentagon official Thomas P. M. Barnett has suggested that the degree to which a country is connected has a large part in defining its place in the modern world, where it stands with respect to the functioning core of nations that participate in globalization and the gap of nations and failed states that have reduced contact with the globalized economy and hence are largely untouched by its benefits. Among the characteristics of the functioning core is the presence of IT infrastructure. Those that must operate in the gap, however, do not have the luxury of having a local Internet service or telecommunications provider to schedule an appointment for the installation and maintenance of service. Satellite services are the mechanism to fill the gap.

For the most part, contemporary U.S. military communication satellite programs are based around geostationary satellites. These are the satellites of the fleet satellite communications system (FLTSATCOM), Defense Satellite Communications System (DSCS), and MILSTAR systems. As in all contemporary ComSats, the primary mission systems of the military ComSats are made up of collections of transponders, the equipment necessary for the satellite to act as an active relay station based at the extreme distance of a geostationary orbit. In addition to bridging the distance of approximately 36,000 kilometers with multiple ground stations simultaneously, military ComSats include provisions for secure communications, which are certain to include encryption and counter jamming capabilities. That being said, the similarities between military and civilian telecommunications have allowed for some off-the-shelf solutions. The much delayed MILSTAR 3, formerly the Advance Extremely High Frequency (AEHF), communication satellite program is based on Lockheed Martin's A2100 satellite bus,[14] which is also in use for several commercial satellites already in orbit. Superficially, all these satellites would bear some physical resemblance; the difference would be found the payload of black boxes that each carries.

Dedicated military satellite communication systems (MILSATCOM) have become victims of their own success relative to the increasing volume of, and demand for, information. Today, reliable and plentiful satellite bandwidth essential to the movement of this information has become an increasing area of concern. Though there is work being done to increase the efficiency of data transmission through data-compression techniques, only more ComSats can ultimately solve current and future requirements. In the meantime, bandwidth (access to satellites in effect) is being purchased from commercial telecommunication companies, such as the restructured Iridium based on the large LEO ComSat constellation concept, and from operators of large traditional GEO ComSats. Of course, the level of security available on commercial ComSats is much less than for MILSATCOM, and military access to commercial capacity competes with nonmilitary users. The expenses of space have meant that ComSats launched for profit are minimally shielded and therefore may be a weak link of space force enhancement.

One consequence of information-age warfare via the rapid movement of information through satellite communications across great distances is the possibility that senior decision makers far away from the battlefield will attempt to micromanage the battle. The challenge is to combine increased connectivity, which technology proponents argue flattens hierarchies, with one of the most hierarchical forms of human interaction—the military chain of command.

As the military adopts information technology, its own lexicon has developed. One of the phrases coined to describe certain transformational concepts partially enabled by space is *network-centric warfare*—a deemphasis on individual platforms in favor of many platforms bound together to work in unison via IT. Just as the business world has seen distances and timescales effectively compress due to the influence of IT, so too has the military world. The reality of warfare today is that observation, decision to engage, and actually conducting the attack on a target does not have to occur on the same platform. Indeed, via communication satellites, these three functions do not have to occur on the same continent. The battlefield of today extends into both cyberspace, with cyberwarfare being a major concern, and into outer space with these critical space lines of communication. Network-centric warfare is more than just enhancing existing weapons, but instead becomes a new methodology for the conduct of military affairs. Among the changes is the creation in a sense of the infinite battlefield, where the battle may be carried out everywhere and anywhere. Compounding the new realties of warfare is the impact of IT on nonmilitary combatants. Mass violence on short notice with little investment is now within the grasp of many actors on the international stage.

Military Space and Three-Block War

It is somewhat misleading to suggest that the military use of space in its force enhancement role is simply the product of technology. Thinking and investments have also been informed by military requirements. As noted in the Introduction, space force enhancement in its initial days was informed by cold war military requirements centered upon the possibility of a major war in Europe, in which the Soviet conventional military superiority forced NATO and the United States to rely on the deterrent threat to use nuclear weapons. The force multiplier effect of marrying the emerging information technologies with space in the 1980s not only offered somewhat of an alternative to nuclear weapons, but also placed space force enhancement within the paradigm of interstate warfare. In so doing, it raised concerns that transforming U.S. military forces to prosecute information warfare was out of touch with the reality of the different form of war or violence in the post–cold war/911 era—the global war on terror and counterinsurgency operations a la Iraq and Afghanistan. Indeed, the chorus of criticisms of the handling of the Iraq War and insurgency by the former secretary of defense Donald Rumsfeld is partially rooted in a failed attempt to apply the information-warfare interstate paradigm to a different form of conflict.

Of course, the information revolution does not just apply to how nations conduct war, but also to how others may use violence to achieve political ends. Unconventional threats such as terrorism and criminal activity have seized greater power through the subversion of information-age technology and other implements of globalization. The terrorist attacks of September 11, 2001, demonstrated how the implements of the modern world can be used by a small group of fanatics to strike not just at the United States, but at the international system as a whole. In light of contemporary terrorism and other substate threats, military transformation and space militarization have had to adapt to remain relevant. The United States and its allies today find themselves embroiled in a global war on terror, which has seen the application of space force enhancement systems against unplanned enemies.

This environment has operational consequences for legitimate militaries designed to fight each other instead of these subnational groups. Terrorists do not generally field armies in the traditional sense, nor sail fleets of warships, nor procure squadrons of aircrafts. Essentially, they do not present targets that were in the sales brochures for military space applications in the 1990s. For U.S. and allied forces on the ground, this presents a complex situation where the level of danger can change in moments. To a large extent, this transformation of war has resulted in threats being found and fought at close quarters. This theater of operation has been described as a *three-block war* where full-scale combat, peacekeeping operations, and humanitarian aid exist simultaneously with only a city block separating each mission. Door-to-door hunts for insurgents and terrorists do not necessarily bring to mind orbiting satellites, yet these systems are part of the joint battle space that also describes today's three-block war.

The proliferation of space systems down to the level of the individual soldier has meant that space is part of the arsenal that the U.S. fields in the global war on terror. Special forces, the tip of the U.S. and allied response to the 9/11 terrorist attacks, entered hostile ground with space systems in tow. As mentioned earlier, the United States had taken steps to prevent satellite imagery from being made available to al-Qaeda and the Taliban during the early war in Afghanistan. Satellite imagery was used for both battlefield intelligence and public relations. Satellite phones allowed easy distribution of intelligence and overall two-way data flow to operators in the field. GPS helped not only with navigation and coordination of allied forces, but also with direct attack. U.S. Special Forces operators embedded with allied Afghan forces could call in satellite-guided bombs. In this respect, space militarization literally rode into war on horseback in Afghanistan.[15] Space systems demonstrated their adaptability and allowed the United States to project information superiority into a land where modern amenities were denied.

The global war on terror is a war of ideas and a war for the hearts and minds of the people in fragile and failed states. Minimizing collateral damage is important in winning such wars. Causing unnecessary harm damages the message that the United States and the international community are a force for good, stability, development, and freedom. Moreover, this is an asymmetric battlefield. The importance of minimizing casualties for the West in these conflicts generally does not apply to the adversaries—terrorists and insurgents. Indeed, maximizing civilian casualties is a tool of terrorists and insurgents to demonstrate the limitations of a government's, or international community's, ability to protect its citizens. Only careful attention to detail and precision can prevent the actions of Western forces being misconstrued into something less than scrupulous. In this regard, the high-tech form of warfare chosen by the United States and its contemporaries is well suited to the global war on terror, as this form of warfare enables a degree of precision necessary to minimize collateral damage. The use of special forces, like the use of precision guided weapons, allows for the surgical application of force. Precision is not only to fulfill the West's preference for a just war but is also cornerstone to victory in what is expected to be a long-term struggle.

Part of winning hearts and minds is found in bridging the gap between the modern world and the nations that thus far have been left behind. Communications satellites allow the information age and modern ideas to be projected into denied parts of the world. The threat posed by communications technology is recognized by certain governments on less than friendly terms with the United States by banning the possession of direct broadcast satellite television receivers and attempts to disrupt these broadcasts through electronic jamming countermeasures.

Conclusion

The emergence of space systems in a force enhancement role is one of the most significant military developments of recent times. It is the backbone in many respects of the information-age warfare paradigm. In this regard, its roots are a function of the development and application of broader information technologies, especially computing, being married to satellites, rather than any major advancements in space technologies related to launch and satellites themselves. Space offers that perch, or the highest of the high ground, from which militaries, led by the United States, can effectively and efficiently exploit these information technologies. Barriers still exist to the full exploitation of space systems in support of terrestrial military operations related to the high costs of space access, on-orbit power generation, and multispectral sensors, among others. Nonetheless, the revolutionary developments of the past two decades suggest that many of these barriers will eventually be overcome.

The significance of space for terrestrial military operations has led many in the United States to describe space today as a critical and vulnerable center of gravity. Potential military adversaries will recognize its critical nature and seek to offset the advantage it provides to U.S. military forces. Adversaries will not only seek to replicate the United States by adapting to information-age warfare, but also undertake strategies and develop capabilities to deny the use of space to U.S. military forces. The measures available to do so range from counterelectronic measures to disrupt the flow of information from and through space to direct attacks on U.S. space assets, military and commercial. In effect, the significance of space force enhancement raises the specter of war being waged in space, and this specter in turns raises demand for developing capabilities to defend space and apply force in and from space.

Notes

1. Global Security, "Historical Nuclear Weapons," http://www.globalsecurity.org/wmd/systems/nuke-list.htm.

2. Ibid.

3. Statements by former Pakistani President Pervez Musharraf and former Indian President A.P.J. Abdul Kalam in 2002 note the deterrent effect of nuclear weapons in preventing open warfare from breaking out recently between India and Pakistan.

4. Fitzsimons, Bernard, ed., et al. "Hs 293, Henschel," *The Illustrated Encyclopedia of 20th Century Weapons and Warfare*, vol. 13 (New York: Columbia House, 1978), 1375–77.

5. Max Boot, *War Made New* (New York: Gotham Books, 2007), 331.

6. U.S. Space Command, *The Long Range Plan* (Colorado Springs, CO: U.S. Space Command, 1997).

7. Norman Polmar and K. J. Moore, *Cold War Submarines* (Washington, DC: Brassey's, 2004), 278.

8. The name *Southern Cross II* is an allusion to the historic 1928 U.S. to Australia flight of the original *Southern Cross*.

9. Dr. Jim Young, Air Force Flight Test Center History Office, "Milestones in Aerospace History at Edwards AFB," August 2007, http://www.af.mil/shared/media/document/AFD-080123-063.pdf.

10. United States Air Force, "Factsheet: JOINT DIRECT ATTACK MUNITION GBU-31/32/38," November 2007, http://www.af.mil/factsheets/factsheet.asp?id=108.

11. Ibid.

12. Thomas L. Friedman, *The World Is Flat: A Brief History of the Twenty-first Century* (New York: Farrar, Straus and Giroux, 2005), 345.

13. Ibid, 350.

14. Peter Bond, *Jane's Space Recognition Guide* (London: Harper Collins, 2008), 216.

15. Lt. Gen. Joseph M. Cosumano Jr., United States Army, "Space Criticality to Ongoing Military Operations," *The Army Space Journal* 1, no. 2 (Spring 2002), http://www.armyspace.army.mil/spacejournal/SJ_V1N2_02_Spring.pdf.

CHAPTER 3

Military Space and Force Application I: Space Surveillance and Passive Measures

For many, the next step of the military use of space is encapsulated by the concept of *force application,* which is generally enwrapped in the long-standing debate on the weaponization of space. Notwithstanding the political, strategic, and economic aspects of this debate, in many senses technology will determine the possible in terms of future space weapons and space warfare. Of course, what is possible will not necessarily translate into reality. As the 1997 U.S. Space Command's *Long Range Plan* emphasized, decisions that translate the possible into reality are political and not military ones.[1] Moreover, decisions to fund certain technologies rather than others are also political. Nonetheless, much of the current debate on force application or weaponization tends to ignore the realm of technological possibility. To understand this debate properly, first one must know what is within reach in the near term.

Predicting the future is littered with many casualties, and space-age predictions are perhaps the poster child of failed dreams. Prediction can be best based upon technologies that are currently being funded. Some of these will reach fruition given continued funding and a reasonable amount of time for maturity. Others may become victims of decisions to terminate funding or may prove to be dead ends. In addition, one must always be sensitive to technological wild cards—technologies that emerge unexpectedly from unexpected sources inside and outside of the military research and development process.

It is also important to recognize in this context the fundamental parameters of force application and the weaponization debate. In general, force application refers to capabilities that enable the application of military force in, to, and from space. Drawing from traditional military concepts, it is generally broken down into three components: space surveillance to provide knowledge about threats to and attacks on space assets, generally conceptualized as *space situational awareness*; passive military offensive and defensive space capabilities, dominated by maneuver and electronic warfare; and active military offensive and defensive military capabilities, which produce kinetic or other destructive effects. While all three are essential elements of force application, the parameters of the weaponization debate are much narrower. This debate is restricted to the active component and largely, but not exclusively, to future active capabilities located in space and on-orbit. As such, it is useful to focus this initial discussion of force application on the first two components and leave the highly contentious weaponization question to a separate discussion in the next chapter.

Space Situational Awareness

Space situational awareness (SSA) is the term used to describe the fusion of sensors and information-processing capabilities critical to the space battlefield. Before one can contemplate defensive or offensive action in space, it is necessary to know what is up there and what everything up there is doing. This is by no means an easy feat. The distances involved, from several hundreds of kilometers to beyond the 35,800 kilometers of GEO, present unique technical challenges. The volume of space that encompasses militarily relevant orbits is magnitudes greater than the volume of airspace over which nations exercise sovereignty. Within this immense volume, objects as small as gloves, loose fasteners, and lost tools must be found and catalogued, as even small misplaced objects given orbital velocities are always potential threats. The importance of space systems to military affairs also includes the additional requirement of differentiating between natural space hazards and hostile actions and intent. The nature of space as a realm of high energy also means that there is little margin for error on matters of attack and defense. Space as a potential battleground does not lend itself to being confined to one small patch or portion of space, because of the nature of orbital dynamics—no object is stationary in space.

The future of space force application will be built on the foundation of SSA as it transitions from basic surveillance into tactical space reconnaissance and targeting. Today, SSA is operationalized in the United States through its Space Surveillance Network (SSN), which consists of radar and optical imaging sensors spread around the world, and in the case of the space-based visible sensor, mounted on satellite. Some of the powerful radars contributing to SSN, such as the Cobra Dane radar in the Aleutian Islands off Alaska, also participate in the U.S. Ballistic Missile Early Warning (BMEW) network, highlighting the similarities between orbiting threats and ballistic missile threats. Other sensors include the Ground-Based Electro-Optical Deep Space Surveillance (GEODSS) telescope complexes spread around the world to watch satellites found at and beyond GEO.

Observations from SSN sensors are combined to provide a picture of what is currently in orbit and the characteristics of each object and its orbit. From these observations, short-term predictions may be made about the future paths of orbiting bodies. SSN performs tracking and orbital predictions for thousands of orbiting objects. Of these, less than a thousand are under human control, with the remaining thousands obeying only the laws of nature and probability. Orbiting objects are subject to numerous forces, meaning that without human intervention, orbits are in continual (though usually minute) change. SSN sensors are only capable of periodic spot checks, providing only a snapshot of the state of orbital traffic. As such, it is necessary to continuously revisit known orbiting objects to update the catalogue of items in orbit.

Most orbiting bodies fall outside the control of U.S. national institutions or for that matter those bound by U.S. law, an excellent reason for continued investment in SSN. One of the primary missions of the SSN is to keep manned space missions, such as the U.S. space shuttle and the International Space Station, and expensive unmanned satellites, such as commercial communication satellites, out of harm's way. As the majority of objects in space are beyond human control, the SSN provides the information to enable human controllers to maneuver assets in order to avoid collisions—space debris or junk has right of way. In the case of the shuttle, the SSN is crucial to ensuring its safety. In addition, when the SSN catalogues a spacecraft capable of collecting intelligence, such as a spy satellite, this information is used to hide sensitive terrestrial activity. For example, during the cold war, highly secretive programs such as U.S. stealth aircraft flights and ballistic missile submarine movements were planned around the schedules for overhead passes by Soviet reconnaissance satellites. With the proliferation of commercial earth imagery, satellites owned by nationals of U.S.-friendly countries must also be taken into account by agencies who wish to keep their activities hidden. Not every commercial satellite is subject to U.S. shutter control regulations—laws that govern observation and information dissemination. Moreover, a nation does not need a highly sophisticated SSN like the United States to be aware of overhead passes of earth observation satellites. Details for the orbits of many satellites, both civil and military, are available in the public domain. For some, collecting satellite observations from backyard telescopes and figuring out the math behind orbits is a hobby.

Things do go wrong in space, which in times of crisis can worsen a situation. Determining the existence or nonexistence of intent in the aftermath of a space incident is important to preventing an unfortunate accident from turning into more serious accusations and worse. Many satellites, such as missile early warning satellites, are considered critical to the defensive stance of world powers such as the United States. The loss of such satellites may be the precursor of an attack. On the other hand, such a loss due to an internal failure or a natural event could just be coincidental during a political crisis. The rapid determination of the cause of a satellite's failure or destruction is important to prevent a critical error. SSN data is already being used to identify historical collisions between orbiting objects. The NASA Orbital Debris Program Office produces the *History of On-Orbit Satellite Fragmentations,* now in its 14th edition, which contains listings for events that are likely to be on-orbit collisions through the study of SSN generated data. Greater coverage to reduce intervals between observations of objects and increased computer power to model events will be required in the future as more and more satellites are placed on-orbit, and military use becomes ever more ubiquitous.

Upgrading SSN is critical to the future of U.S. space power. Even without the general deployment of space weapons, SSN will have more to do. The expansion of military and civilian/commercial space capabilities among nations continues to grow. More activity in space means more objects and more debris on-orbit, and consequently more opportunities for catastrophic on-orbit collisions. It is even theorized that LEO may be rendered unusable by a runaway chain reaction where space debris generated by satellite collision rapidly leads to more satellite collisions, generating more debris, which in turn continues the chain. This scenario, dubbed the Kessler syndrome, after the NASA consultant who proposed it, Donald J. Kessler, only becomes possible if there are enough orbiting bodies. This orbital catastrophe is somewhat akin to the self-sustaining runaway fission reaction of an atomic bomb being dependent on having a sufficient critical mass of fissile material. The critical mass of orbiting bodies needed for the Kessler syndrome to become reality is at present a matter of scientific debate. However, with the number of objects being launched into orbit exceeding the number deorbiting, this may become a real problem in future.

In order for the United States to be able to counter the military space systems or force application capabilities of potential adversaries, SSA is crucial. As discussed in the last chapter, space force enhancement is an integral component of the information-age warfare paradigm and therefore something for competitors to both emulate and counter. The U.S. example is being followed—competitors and allies are racing to produce equivalents to systems such as U.S. intelligence gathering and GPS satellites. The existence of these advanced systems must be taken into account for national security operations, even in peacetime. While the United States cannot stop other nations from developing and deploying their own equivalents of space force enhancement systems, the United States must remain aware of foreign military activities in space.

The global security situation is dynamic, and only through sustainment and enhancement of the science and technology of space surveillance can the United States maintain space superiority. Ground-based radars, primarily dedicated to ballistic missile tracking and ballistic missile defense, already play a significant role in space surveillance and U.S. space superiority. In the future, the planned Space Surveillance and Tracking System (STSS), especially the planned infrared sensor satellite constellation of 24 satellites in LEO, formerly known as space-based infrared-low (SBIRS-L), designed to provide launch detection, and missile defense tracking, cueing, and target discrimination during the midcourse phase of the ballistic missile bus or warhead, will also serve a vital SSA role, as well as a force enhancement one. Another facet of the ballistic missile/missile defense dynamic that serves an SSA role is launch detection and identification. Missile early warning is done presently through the U.S. Defense Support Program (DSP) of infrared sensors in geosynchronous orbit, which is being modernized with a new generation of sensors as part of the SSTS, the space-based infrared-high (SBIRS-H). According to public accounts, SBIRS-H will consist of four satellites in GEO and two satellites in a high earth orbit (HEO) to provide coverage of the north, with the first launched into HEO on June 28, 2006.[2]

The future of SSA is to move sensors directly into space. Leveraging more directly off of the missile defense effort is the space-based visible (SBV) sensor, mounted on the Midcourse Space Experiment (MSX) satellite, which was launched in 1996 primarily to study how a ballistic missile in flight could be detected and tracked. After over a decade in orbit, and long after the missile defense experiments ended, the SBV sensor continues to provide practical observations to the U.S. SSN. In the wake of the success of SBV are plans for a satellite constellation dedicated to space surveillance, the Space-Based Space Surveillance (SBSS) program. Just as the three ground based GEODSS facilities are the satellite surveillance counterparts to the numerous astronomic research observatories, sensors like SBV and the future SBSS constellation are the satellite observation counterparts to the Hubble space telescope. From space it is expected that superior observations can be made than those from the ground with little regard for atmospheric weather and no need to wait hours for night to begin operation. The prototype for this system is the SBSS Pathfinder, which is expected to be launched in 2009.

Two events highlighting the importance of SSA and the limitations of the current U.S. and international SSA capabilities gained some attention in early 2009. On February 2, 2009, Iran first successfully lofted an operational satellite, the *Omid* (Persian for "Hope") via its own launch vehicle, the *Safir-2*. Aside from the ICBM potential that any orbital launch vehicle technology implies, Iran's development of an orbital space launch capability demonstrates that space technology is proliferating, including to countries with troubled relations with the West. As space is directly accessed by more nations, it will be critical to have robust space monitoring, as it is unlikely that future Iranian satellites will be just for "telecommunication and research."[3] Later on, in February 2009, an active Iridium LEO communications satellite collided destructively with a defunct Russian satellite. Now while the collision between Iridium 33 and Kosmos 2251 on February 11, 2009,[4] was an accident, it is perhaps more worrying than the Iranian launch earlier in the month. Iridium 33 and Kosmos 2251 are not small pieces of space debris. Both were intact satellites. The fact that they did slip though the cracks of the U.S. and other space surveillance and crash in orbit only points to the limits of today's capabilities. It is compounded by the resulting debris clouds, which are now a hazard to many satellites, including the other satellites of the large Iridium constellation. Iran's space activities might actually be as nonthreatening as Iran claims, as with other space technology developments made outside the West. However, the odds of additional space collisions involving operational satellites only grows with the increase in space activity, making vigilance over what is happening overhead all the more critical.

Another form of SSA is the detection of communications interference. The Rapid Attack Identification Detection Reporting System (RAIDRS), soon to reach full operating capability, is to provide timely detection, characterization, and location of sources of electromagnetic interference that may be affecting communications with fielded forces.[5] A satellite is as good as destroyed in orbit if it cannot deliver services to those on the ground. In providing information on such interference, RAIDRS enables action to be taken against the sources of interference, ranging from counter jamming measures to bypass or otherwise burn through interference to actual attack against them.

The 2001 Space Commission chaired by Donald Rumsfeld warned of a space Pearl Harbor. It identified potential threats to U.S. space power, ranging from the usual competing nations to subnational groups such as terrorists. Pearl Harbor is a metaphor for being caught unaware. Preventing the space equivalent of a Pearl Harbor is to prevent being caught unaware of an attack on U.S. assets in space, and SSA is the cornerstone to preventing such an outcome.

Maneuver and Stealth

Beyond the challenges of detecting and tracking objects in the vastness of space is the tactical use of the very vastness of space. The distances between the surface of the earth and potential orbiting targets provide some measure of protection. In many respects, maneuver to avoid an interception is easier than the effort needed to successfully carry out an attack. The problem of keeping track of thousands of space objects is also an opportunity for one's own satellites to get lost among the clutter. The longer a satellite remains in a particular orbit, the greater the opportunity for observation and hence prediction of its future location. Thus, a satellite that is both free to move and hard to detect has a capacity to defeat interception attempts.

Space is a realm of high energies, but rarely does this energy lead to rapid and spectacular maneuvering. The space dogfight, a staple of science fiction, requires technologies still far from reality. The momentum, or great investment in energy, that keeps a satellite from falling from the sky also limits on-orbit maneuvering. Any thrust generated by a satellite will result in a change in the orbit, but as the capacity for generating thrust onboard a satellite is relatively small compared to the thrust capability needed to put a satellite into a stable orbit, orbital changes are similarly small. Satellites do not generally carry the propulsive capability to make rapid orbit changes. Unless a spacecraft is meant to carry out rapid orbital changes, its internal propulsion capability is kept to a minimum. This has thus far limited maneuvering to countering primarily the small, but always present, natural forces that shift a satellite's orbit.

On-orbit maneuver is a problem of how to best make use of the limited mass and volume allowances permitted by contemporary launch vehicles and budgets. Most propulsion systems require the expenditure of propellant; therefore, increasing the efficiency of an engine's propellant use would intuitively increase opportunities for maneuver. Rocket efficiency is in part defined by how much velocity can be imparted on reaction mass—the greater the exhaust velocity of a rocket, the greater the efficiency of the rocket. Chemical rockets use the propellant as both the energy source and the reaction mass, whereby a chemical reaction generates energy, which is then harnessed to eject the by-products of the reaction as reaction mass. The exhaust products of the chemical reactions that power chemical rockets tend to have relatively high molecular mass, meaning that more energy is associated with accelerating these molecules to a particular velocity. Storable chemical propellants, those necessary for satellites with missions measured in years, tend to involve larger, more complex molecules. Consequently, on-orbit propulsion tends to be less efficient than launch vehicle rocket engines such as those using simple hydrogen and oxygen propellants. Moreover, storable chemical reactants have a tendency to be less energetic than cryogenic fuels like hydrogen and oxygen. One method to increase on-orbit propulsion efficiency, and allow greater freedom of movement, is to abandon chemical rocket propulsion.

Electric rocket engines are now being promoted as a way to escape the inefficiencies of storable propellant chemical rockets for on-orbit maneuvering. In an electric rocket, the reaction mass is accelerated to higher velocities from an external energy source. This energy can be easily obtained from the sun while in orbit, though heavy energy storage methods are still needed when a spacecraft is in the dark. The sun as an energy source in orbit has the major advantage of being practically limitless and free of operational costs (startup costs for solar power are an entirely different matter). It also helps that the leading solar energy conversion technology, photovoltaic cells, are solid state, relatively light and have long life spans. Alternatively, nuclear power can also provide continuous high levels of electricity, though developments in converting the sun's power into useful electricity have reduced the attractiveness of this energy source, along with existing political concerns about nuclear power.

Electrical energy can be used to accelerate small quantities of propellant at very high velocities. Thermal rockets, which rapidly heat reaction mass so it expands out a nozzle, can be powered by electrically powered heating elements, or by arcing in the same manner as a spark plug. The limits of a thermal rocket are defined by how much energy can be delivered to a reaction material and the ability of the engine material to withstand the resulting temperatures. Thermal rockets may of course use other sources of heat, with nuclear and direct solar having gained serious attention. Given that energy may be plentiful with nuclear or solar sources of energy, very high exhaust velocities are achievable in thermal rockets. Practically any reaction mass that can be ejected by temperatures within the tolerances of the rocket engine may be used. However, considerations over storage and low molecular weight determine what is actually used.

Electrical and magnetic fields can also be used to accelerate reaction mass to comparatively high velocities. The *ion-drive* concept popular in science fiction is based on the very real science of electric fields. Real-world ion rocket engines are capable of very efficient use of propellant, by virtue of high exhaust velocity, but are capable of only low amounts of thrust. This means that a contemporary ion engine can efficiently change the velocity of a spacecraft, but only over a very long time. For station-keeping maneuvers, the thrust limitations of ion rocket thrusters are no problem as these engine firings are meant to counter usually more subtle natural forces, and they have already found use on long-duration orbital satellites and to propel a growing number of deep-space probes. However, the low thrust potential of ion propulsion limits its usefulness in attack avoidance maneuvers.

Development continues into increasing thrust levels from electric rockets. One promising technology is retired NASA astronaut Dr. Franklin Chang-Diaz's variable specific impulse magnetoplasma rocket (VASIMR). The VASIMR concept is envisioned as being able to tailor specific impulse levels for high thrust at the expense of efficiency, or high efficiency at the expense of thrust levels. Avoiding the complexity and risks of an untried dual-mode electric engine, dual engines of proven lineage could be used. A military satellite could be equipped with a storable inefficient high-thrust engine, such as a small solid rocket motor, meant for emergency maneuvering, while using efficient low-thrust propulsion for station keeping and returning the satellite to a useful orbit when out of danger.

Maneuver does not necessarily mean expending a consumable. Outer space is an environment that for all practical definitions is without the dampening effects of friction. Newton's law of action and reaction does not just apply to rockets. Spinning a wheel onboard a spacecraft will result in a rotational reactive force for the satellite around the wheel's axis. In other words, the satellite generates an equal but opposite spin force to the wheel, otherwise known as torque. This effect is seen on earth in the ordinary helicopter where the transmission of rotational force to spin a helicopter's big lift generating rotor must be counteracted by a smaller tail rotor or having a second big lift rotor spinning in the opposite direction, or other aerodynamic systems such as MD Helicopters' NO Tail Rotor (NOTAR) system. Electrically powered momentum and reaction wheels can stabilize and even perform fine adjustment of a spacecraft's orientation, its attitude, without any consuming anything, unless one counts the wear and tear on these torque-generating devices over many years of service.

Another means of propellantless on-orbit maneuver is via an electrodynamic tether. A conductive tether (rope) oriented across the earth's magnetic field can generate changes to a spacecraft's orbital velocity and consequently altitude. Running an electric current across a magnetic field, whether it is the earth's magnetic field or the magnet of an electric motor, will result in force being generated. In an electrodynamic tether, this force has the net result of raising the orbit of the tether and the satellite to which it is attached. Moving a conductor (as part of an electric circuit) perpendicular to a magnetic field will convert motion into electricity. This is the same principle behind an electric generator and when applied to an electrodynamic tether will convert the motion of a satellite into electricity, with the net result of lowering its orbit. While the theory behind the electrodynamic tether is simple, the engineering is somewhat difficult as it means producing and deploying a conductive cable enclosed within an insulated sheath that is several kilometers long. This is a technology that has been test flown numerous times but has yet to be used operationally.

Light itself can be harnessed to provide thrust without propellant. Light exerts a small but real amount of force, which can be harnessed given enough surface area. Solar sails represent a possible solution for on-orbit propulsion; however, they have limited applications to satellite defensive and offensive maneuvers. For one thing, solar sales would interfere with the militarily desirable characteristic of being inconspicuous.

Mitigating the problem of limited consumable supplies are various concepts to refuel satellites in orbit. Conceptually, this is an extension of aerial refueling—a capability that gives the U.S. Air Force the reach to perform global strikes. Fuel and even modular spare parts may be lofted for much less cost than complete satellites. At the very least, the insurance costs of launching a cargo of fuel into a parking orbit for later retrieval and use would be lower than that for a full-up unitary mission satellite.

The U.S. space shuttle has already demonstrated the basic ability to service satellites on-orbit. The Hubble space telescope (HST) has been serviced by five shuttle flights since its launch in April of 1990, with the last occurring in May 2009 and entailing a major overhaul to the telescope. The first servicing mission had the task of adding equipment to correct for a manufacturer's defect in the optics, which allowed the HST to finally meet the great expectations put on this program. Subsequent servicing missions have allowed the HST to continue on its mission of discovery for nearly two decades, through replacement of batteries, solar panels, and gyroscopes. The capabilities of the HST have also been improved through upgrades to the instruments. The problem with using the space shuttle to service satellites is that all shuttle missions are expensive and are too risky to be considered routine even in the military sense. In many ways, the costs associated with the reusable space shuttle is justification for making spacecraft disposable. Instead of being retrieved, an idea that was being entertained before the loss of the space shuttle *Columbia,* the fifth and final servicing mission was to attach equipment to facilitate a final rendezvous, this time by an autonomous spacecraft that will attach propulsion to perform a controlled deorbit of the HST from LEO.

The servicing history of the HST points the way toward the future of on-orbit support, from manned missions eventually to a final robotic mission to decommission this important scientific instrument. The Orbital Express program run by the Defense Advanced Research Projects Agency (DARPA) with the cooperation of NASA has already demonstrated many of the capabilities needed for robotic servicing of an orbiting satellite. Two satellites, the Autonomous Space Transport Robotic Operations (ASTRO) and Next Generation Satellite/Commodity Spacecraft (NextSat/CSC) were lofted into orbit in March of 2007. In orbit and separated, these two satellites autonomously performed rendezvous and finally capture/docking maneuvers. Once connected, propellant was transferred between the two satellites. In addition, the robotic-arm-equipped ASTRO was able to swap components with the NextSat/CSC. These two small satellites were able to perform these tasks with a minimum of human control. The success of Orbital Express paves the way for greater freedom to maneuver by future military satellites, not to overlook the military implications of a capability to capture a satellite. Advanced servicing satellites developed from ASTRO will be able to rendezvous with low-cost bulk cargo (commodities) spacecraft equipped with minimal systems, load up on fuel and parts, and then maneuver to link up with client mission satellites to top up their propellant reserves and even swap out equipment.

The future of satellites being able to refuel other satellites requires the development of a common architecture to define accessories for docking, refueling, and hardware interfaces. This is no easy feat, as in the world of aerial refueling the United States has ended up with two systems of aerial refueling, the boom and receptacle system that is used by the U.S. Air Force, and the probe and drogue used by the U.S. Navy (and most of the United States' allies). This is less a technical problem than a bureaucratic one, as the raw knowledge of how to accomplish on-orbit refueling has been demonstrated.

Dispensing with the problems of having to design from the ground up satellites for on-orbit servicing, the European Space Agency has proposed the ConeXpress Orbital Life Extension Vehicle (ConeXpress OLEV) small satellite. At its heart, ConeXpress is a propulsion package with the capability to take over the maneuvering needs of a large satellite. To accomplish the mission of replacing the propulsion capability of a dry satellite, the ConeXpress system uses efficient Hall effect electric thrusters[6] and a system to clamp onto the nozzle of the relatively large rocket engine found attached to most GEO satellites used for final orbital placement. Once attached, the ConeXpress would act as a tugboat, replacing the onboard propulsion for station keeping and other necessary maneuvers to keep the client satellite useful.

The backers of the ConeXpress OLEV envision this system as a way to add up to 10 additional years to the life of a satellite or, alternatively, salvage satellites that are stranded in a useless orbit due to launch mishap. ConeXpress OLEV is also promoted for decommissioning satellites by flying them up to high graveyard orbits or forcing them into the atmosphere in a controlled manner. In possessing this ability to move satellites, ConeXpress has potential military application, even though it is being offered as a commercial service. In fact, any capability designed to service an on-orbit satellite has offensive and defensive military application, including the U.S. space shuttle and its robotic CanadArm.

Maneuver is only one countermeasure to attack. The other is to make it intentionally difficult for others to actually observe the satellite. Space launch is a spectacular event, which produces a wealth of data from which a satellite's initial orbit may be inferred. Once in a stable orbit, the lower power propulsion systems used for orbital maneuvering allow for subtle changes to orbit, obscuring the final mission orbit of the satellite. This effort to mask over the mission orbit of the satellite would go to waste if the satellite were easily detected and its new orbit recalculated. Low observability (LO) technology, otherwise known as stealth technology, is the other half of making a satellite harder to hit. For every signature that an object may be detected and tracked with, there is a corresponding discipline devoted to reducing it to imperceptible levels. The world of LO technology development is for obvious reasons cloaked in secrecy. Scraps of publicly available information on satellite stealth are rare, and in all likelihood just a vague foreshadowing of what will only be declassified many years in the future, as evident with the 1960s CORONA Project. Satellite tracking is performed primarily optically and via very powerful radars, which means that satellite stealth technology likely concentrates on reducing optical and radar signatures.

Optical camouflage, the art of making a satellite appear to be something else, has long been a concern in space security matters. From the earth, a satellite to the naked eye and most optical telescopes appears to be sparkling points of light racing across the sky. The atmosphere, however, with its every changing density and composition distorts imagery. Moreover, low-orbiting satellites move rapidly across the sky relative to an observer on the surface of the earth, presenting extremely short windows of opportunity for a picture to be captured. Most high-power telescopes can only produce a blurry smudge of light when pointed at satellites. Just what a particular foreign blur is doing in orbit is often unknown.

Satellites, unlike military aircraft, take on rather generic and utilitarian shapes, which often do not disclose their purpose. Concealing a weapon or other military payload is as simple as enclosing the offending system within a lightweight false structure. The precedence for such concealment is found in the Q-Ships of World War I and II, where antisubmarine warships were disguised as vulnerable merchantmen with false structures and sometimes going as far as flying false flags to enhance the deception. While the practice of false flags is somewhat questionable diplomatically, there are and never have been norms and conventions against orbiting spacecraft under false pretenses. It is known now that some Soviet civilian space stations of the Salyut series were actually military space stations of the Almaz program, which in addition to reconnaissance payloads were armed.[7]

The difficulty in obtaining detailed imagery on foreign satellites in orbit has fueled fears over just what is overhead. Discerning the intentions of foreign satellites is something of an imprecise art, involving comparing orbits with those of known satellites or those expected and surreptitiously overhearing or monitoring satellite telemetry. During the cold war, for example, monitoring telemetry was an essential means used by both the United States and the Soviet Union to promote stability through the agreed use of national technical means in support of the various strategic arms control agreements, which began with the first Strategic Arms Limitation Treaty (SALT 1) in 1972. More down-to-earth forms of intelligence are also used, such as when U.S. reconnaissance satellites were compromised by William Kampiles, who sold the manual to the Soviet Union.

One of the best-known facilities for obtaining sharp pictures of space objects is the Starfire Optical Range (SOR) located at Kirkland Air Force Base in New Mexico. Using adaptive optics technology to compensate for atmospheric distortion and precise mechanical telescope pointing, SOR is able to take relatively detailed imagery of LEO satellites.[8] Recently, debate occurred over whether the Starfire telescopes could have detected the fatal damage sustained by the space shuttle *Columbia* during launch. Adding to the debate is a well-circulated image purported to show the damaged left wing taken by engineers at Kirkland AFB. The picture in question was not taken by one of the SOR telescopes, but instead via an apparatus put together as a spare time hobby by the engineers, which consisted of an amateur astronomer's telescope given computer control by an obsolete home computer.[9] The photo was a demonstration of the human capital at the SOR facility rather than any military hardware.

The orbiting SBV sensor and the follow-on SBSS constellation of satellites make claim to superior satellite observation capabilities compared to ground based telescopes in part by avoiding the problems of atmospheric distortion. The clarity of postrepair Hubble space telescope imagery attests to the benefits of being above the denser portions of the atmosphere when it comes to imagery. The adaptive optics technology being pioneered at the SOR and elsewhere, in addition to developments in field of computer image enhancement, may allow ground-based telescopes to compete against their more expensive orbiting counterparts in both astronomy and national security applications. In all likelihood, the future will see a mix of both ground-based and space-based telescopes for both research into the secrets of the universe and the secrets of orbiting satellites.

The ease at which a satellites true purpose may be camouflaged remains a critical argument against space weapons treaties. For a treaty to be effective the parties involved must have reassurance that the other parties are not secretly breaking the treaty. As the purpose of a satellite can be easily disguised, it makes treaty verification difficult if not impossible. If treaty adherence by an international partner cannot be reliably verified, then it is of questionable interest for the United States, and other states, to pursue a treaty, notwithstanding the political forces at play. In the real world, it is very difficult to keep secrets in open and transparent societies such as the United States, where trade and even popular press publications are able to speculate on secret programs inferred from publicly available documents. Moreover lively open debate over the legality of specific weapon systems with respect to treaty obligations not only exists, but is also encouraged as part of democratic society. This debate even encompasses holding systems to higher standards than what is actually specified in treaty, such as the continued fallacy that the 1967 Outer Space Treaty bans all weapons in space and not just weapons of mass destruction (WMD) in orbit. Closed societies on the other hand, such as was found in the Soviet Union, and in varying degrees in Russia and China, do not permit such speculation. As a result, Western nations are at a disadvantage when entering into such treaties due to the sheer physical difficulty of verifying treaty compliance.

Avoiding detection by radar is most commonly associated with stealth technology. Radar cross-section (RCS) reduction is a combination of having the correct shape, materials, and understanding of radar systems. Ideas concerning low radar observability date back to World War II, when it was noted that certain materials and certain shapes were only detected on radar at shorter ranges compared to others. The flying wing configuration first proposed for its aerodynamic efficiency and now used in the U.S. Air Force's B-2A Spirit Stealth Bombers, subsequently demonstrated radar stealth qualities to exploit this configuration such as Northrop's earlier B-35 and the Horten brothers' flying wings during World War II.

Calculations on how to predict radar returns where developed and published in Russia by Pyotr Ya Ufimtsev and were used by Lockheed Martin's Skunk Works among others in the 1970s for computer programs designed to predict radar returns. These early computer models for RCS led to the first true stealth aircraft, Lockheed's Have Blue test bed and the operational F-117 attack aircraft.

RCS reduction technology is not just confined to aviation, and has been applied to other weapons platforms most notably warships, with RCS techniques defining the shape of Sweden's Visby class corvettes, France's La Fayette and is currently being applied to the U.S. Navy's Gerald Ford class of aircraft carrier now under construction. Like space, naval applications have little worry about aerodynamics, giving a degree of freedom for low radar return design. Space applications for RCS reduction are certain to have been proposed, though such developments would combine the two most secretive worlds—military and intelligence space programs, and the world of stealth.

Stealth technology can be applied in varying degrees to lower a platform's signature. The greater the desired the level of signature reduction the greater the cost. For this reason, many applications of stealth technology are aimed only at reducing a platform's signature to stall detection and concentrate on specific views or aspects of the platform. At present most satellite observation is done from the earth, and therefore it is expected that satellite stealth would be primarily applied to the earth facing side of a satellite. For the time being, all aspect stealth found on aircraft such as the B-2A would seem to be unnecessary for satellite applications, reducing the problems and costs associated with stealth in orbit. Another rationale for only having stealth applied in the direction facing earth is that there are all sorts of projections such as solar panels and radiators that must be exposed to space to function; otherwise there is the problem of building a stealthy solar panel or radiator.

There is also speculation that satellite stealth takes the form of a deployable screen, which shields the satellite from view of the earth. This barrier would be shaped and composed to deflect and/or absorb radar wavelengths used for satellite detection and tracking. In addition, measures would be taken to minimize reflecting light back toward the earth, such as pigmentation and even active systems that match the illumination with ambient lighting conditions. A screen that is big enough to hide all of the not-so-stealthy parts common to a satellite would be relatively large, requiring that the screen be deployable once the confines of the launch vehicle are escaped. The idea of a deployable stealth shield for satellites to hide behind is not new. U.S. Patent 5,345,238 concerns just such a system and was issued to Teledyne in 1994.

Some also speculate that stealth satellites are not the future of U.S. space power, but instead are already part of the present. Numerous articles in respected aviation and technology media cite a stealth reconnaissance satellite program code-named MISTY and even orbited satellites attributed to this program such as AFP-731 (otherwise known as satellite US 51). Broadly, the MISTY satellites are claimed to combine the digital camera based optical reconnaissance capabilities of the relatively well known KH-11 KENNAN/CRYSTAL satellites with radar and optical deception technology, such as the deployable stealth screens described earlier. What is known is that on a few occasions, satellites such as AFP-731/US 51, which were easily observed and documented by amateur satellite trackers to be in orbits useful for reconnaissance, have ceased to be easily observable. AFP-731/US 51 is a recognized U.S. satellite payload, and at the time of writing, even had an entry in the National Space Science Data Center, as NSSDC ID: 1990–019B.[10]

Stealth must be tailored toward the expected opposing sensors. Earth-based optical and radar sensors are the primary methods for space tracking in use today and therefore today's satellite stealth will concentrate on defeating these two sensors. Historically, during the cold war, another form of space observation was in vogue, telemetry ease dropping. Once unconvincing Soviet fishing trawlers sailed international waters near Western space facilities in hopes of gathering some tidbits on what was being launched from Vandenberg or Cape Canaveral. Satellite interlink communications allow satellites to obscure their communications activity by not directly linking to ground stations. Radio direction finding, the ability to detect and localize communications signals, may still be used to detect, track, and categorize satellites, but like most things related to signals intelligence, it would be shrouded by secrecy. It is certainly not noted as being part of today's U.S. Space Surveillance Network. In future the increased use of space-based sensors for satellite observation will change the definition for what is low observability for a satellite. Space-based sensors will increase the directions from which a satellite may be observed. Hiding a satellite behind the cloak of a simple screen will certainly not be enough. This once again is a problem of how much stealth the United States or another nation is willing to pay for.

Damage Limitation

Sooner or later, dodging and hiding will fail to protect a satellite from attack, such that the next order of business is to mitigate damage. In some senses, satellites are under continuous attack from nature, and many satellite protection schemes may be inferred from known spacecraft hardening techniques. For some destructive mechanisms, the only difference between attack and a natural hazard is a matter of intent. There is also the option of not worrying about extra shielding and making up for the satellite loss in some other way. In effects-based warfare, it is not necessarily that the satellite be preserved per se, but that the capabilities it provides be ensured. Finally, there is the option to threaten to respond to an attack disproportionately. A satellite attack will not happen in a political vacuum, and making certain that potential foes knows the possible repercussions for such an action will potentially deter such an attack from being considered.

A satellite in orbit is continually bathed in radiation, which given time will destroy its computers. To forestall this eventuality, measures are taken to allow electronics to withstand radiation conditions expected to be encountered by a satellite during its life span. Present-day computer technology is based primarily on digital logic (on or off, yes or no, 1 or 0) processed in semiconductor integrated circuits. In semiconductor-based logic circuits, data is stored and processed by the presence or absence of an electrical charge. Radiation and other electromagnetic interference can disturb the electrical state in a semiconductor device, changing ones to zeros, or vice versa, corrupting data. Permanent damage results when radiation changes the critical material properties that allow a circuit to function and by the electrical effects generated by high-energy particles passing through components.

Radiation hardening involves a combination of hardware techniques such as shielding against radiation effects and redundancies, and through software error correction meant to catch and correct any data corruption that occurs. Passive shielding, armor, or *hardening*—the term applied in the nuclear weapons sphere—prevents the entry of radiation and electromagnetic interference. Military systems have a tendency to be hardened for the extremely heavy doses of radiation and electromagnetic interference generated by nuclear warfare. However, not withstanding the amount of bulk proposed in the Project Orion nuclear bomb propulsion system, there are severe limits to how much nuclear shielding can be lofted with today's space launchers. Moreover, it also adds dead weight to a satellite, thereby increasing launch costs or reducing the payload, and, in general, is avoided in civilian/commercial applications.

Redundancies, spare parts already in place and immediately ready to take over, can allow for a system to degrade gracefully as elements fail. Software fault correction by itself does not add to the mass of the satellite. Ongoing advances in data storage have given plenty of room for programs to grow bigger, and satellite software does not suffer as much from the bloating that seems endemic to consumer-level software. However, software development and testing are now large parts of the total cost of ownership of many aerospace and military products. It is of note that vacuum tubes and older technology that was used in digital computing, which predates semiconductors and even the solid state transistor, are much more survivable with respect to radiation and EM interference. Vacuum tubes, which are similar in size and construction to light bulbs, are, however, much larger than the microscopic logic gates found in the average computer chip.

The weight limitations that affect the amount of electronic shielding also limits protection against physical impact. In space the majority of physical impacts come from small debris, both natural and artificial. Short term space shuttle missions have on occasion returned from orbit with pits and scratches from high-speed impacts from objects too small to be visible to current U.S. space surveillance. The whipple bumper, a type of shielding composed of one or more thin shields separated by empty space from a final thick layer of shielding, is one possible method to get around the mass problems of armoring against impact. The energy of a hypervelocity impact vaporizes the object impacting the whipple bumper shield, reducing it to a harmless shower of droplets unable to penetrate the final thick layer of spacecraft shielding. Whipple bumpers are unable to stop large objects but may be able to offer protection from intentional attacks composed of smaller debris being thrown into the path of a spacecraft.

Shielding satellites from attack is arguably a futile affair. Like armor on earth, armor in space can be defeated by simply increasing the strength of an attack. The growth margins for hardening spacecraft against attack are much smaller than that possible for tanks and certainly less than that available to warships. Every spacecraft is a compromise between mission payload and common supporting systems of the satellite bus. Armor only adds to pressures on this already demanding balance. Accepting that an antisatellite attack is in all likelihood not survivable does not mean giving up on space power.

Today there is great interest in opening up LEO through not only cheaper space access, but rapid or on-demand space access. In the civilian world, cheap on-demand access to space is part of the drive to open suborbital and later orbital flight to tourists. For military space power, cheap on-demand space access will allow satellite replacement should warfare reach orbit. Rapid replacement of satellites will mitigate some of their inherent vulnerabilities. The concept of responsive space access is part of the umbrella of operational responsive space (ORS), which aims to make space power a more agile asset, able to respond to new demands in a timely manner, which includes the ability to react to an attack.

Like *military transformation,* ORS is one of those phrases that can mean different things depending on the context and audience. For some, ORS is a new way of conducting national security space programs, leveraging concepts from the computer industry, such as plug-and-play modular components, to produce building blocks for satellites that can be rapidly assembled into small satellites and launched to provide new or replacement capabilities as needed. This is in contrast to the current model for military and other national security space missions employing large, often multimission, satellites, which have a tendency to be expensive and require years to get from concept to orbit. The goal of ORS is to create an infrastructure able to generate rapidly both new and replacement space capabilities for warfighters.

Many factors such as advances in computer technology and small satellite propulsion have allowed existing and improved capabilities to be fitted into a smaller satellite. However, smaller satellites cannot be necessarily hardened as much as their larger counterparts. Nonetheless, small satellites offer other methods to avoid and mitigate the effects of an attack. Small satellites are inherently more difficult to find and track. Launch costs are potentially lower for small satellites and allow the use of smaller, less-sophisticated launchers. Instead of satellite programs that resemble the acquisition program for a major warship, small satellites can be churned out like munitions or disposable sonar buoys. Switching from the traditional large satellite military model of practically irreplaceable satellites to swarms of expendable short life time satellites with rapid replacement rates would make destroying U.S. space power capabilities a difficult proposition.

Smaller satellites do not necessarily mean loss of capability. In many high-technology applications, large monolithic systems are being replaced by arrays of smaller units that work together to produce the same or superior results, often at lower overall price and with greater speed. Examples of parallel computing have problems broken down and spread across dozens, if not hundreds, of individual computers. In radio astronomy, arrays made up of dozens of radio telescopes working in unison can produce observations more in line with dishes of impractically large size. The Very Large Array (VLA) radio telescope in New Mexico is made up of 27 ordinary size radio telescopes to produce the effect of a dish 36 kilometers in diameter.[11] Signals intelligence for instance is in many respects simply radio astronomy pointed toward the earth instead of the sky. Active electronically scanned array (AESA) radars, which are now in vogue for the newest warships and warplanes, are based on multiple small relatively low power electronic modules each capable of generating, transmitting, and receiving its own radar beam. AESA radars are also touted for their intelligence-gathering abilities. In both radio telescopes and AESA radars, the results from these smaller modules are combined. Network-centric warfare concepts have sensors distributed across a battlefield, with the observations made by individual platforms combined to produce the effect of information superiority. Small satellite arrays may not meet the requirements of every space power application, but earthbound counterparts have already proved the concepts.

Physical separation between satellites is not necessarily a problem, as computer networking has made the distances between elements largely irrelevant. High-speed communications between satellites is of course a less-developed technology than earthly networking, where there is plenty of room for bulky equipment and redundancies. However, such capabilities are not in the realm of science fiction. The Iridium constellation in service since the 1990s uses intersatellite communication links to route data between satellites so that a phone call made through the Iridium service from a handheld phone can pass through multiple satellites before retuning to earth-based downlink station, or even to another Iridium handset.

Physical separation between satellites in an array is another a form of protection. Proposals exist to replace the function of a single, large satellite with clusters of small identical or compatible satellites. In such a situation, the loss of an individual satellite would only degrade the capability of the formation, not destroy it outright. Given a big enough cluster of satellites, the loss of a few satellites could be easily shrugged off through automatic rearrangement of satellites to make up for gaps or by providing on-orbit redundant satellites that can be switched on when needed to replace losses. Autonomous satellite control and formation flying would be necessary in conjunction with low-cost launch for small satellites to make such clusters of satellites feasible.

Responsive space, the ability to quickly put together a satellite or satellite cluster to carry out a newly identified space power requirement, is also the ability to rapidly replace a space power capability. In a way a responsive launch capability has been around for decades in the form of the ICBM. Many early orbital launch vehicles are based on ICBMs and in many cases are identical with the exception of flight path and payload. Today's reactive space launch faces a different, less generous, more competitive military spending climate. That being said, small launch vehicles with short response times are being funded as part of DARPA's Force Application and Launch from Continental United States (FALCON) program. FALCON's ultimate goal is the development of high-speed global payload delivery capability from the continental United States (essentially a very long-range and fast bomber), with both suborbital reentry vehicle and air-breathing hypersonic aircraft approaches being investigated. In support of these long-term goals is FALCON's small launch vehicle (SLV)—the development of a modest launcher for both testing hardware developed for FALCON's hypersonic vehicle branches and for launching small satellites cheaply. Alongside aerospace giant Lockheed, those vying for SLV contracts include small businesses such as AirLaunch LLC, Microcosm, and SpaceX. Of the competing launch vehicles, SpaceX's Falcon I as of September 2008 has attempted four orbital launches, with each one providing valuable operational data for low-cost and rapid space access.

FALCON is just the latest in a long history of U.S. hypersonic aviation development. Some programs such as the Reagan era X-30 reached far but fell short, ultimately being cancelled. The less ambitious X-43 Hyper-X, an unmanned research aircraft recently demonstrated the near-term viability of the supersonic combustion ramjet (scramjet)—a jet engine meant to operate at hypersonic speeds which figured prominently in the earlier X-30 program. Only time will tell if FALCON's somewhat less-ambitious goals are achievable on schedule and on budget.

Another long-term military space access capability is the military space plane. U.S. interest in a military space plane includes the X-20 Dynasoar program, which was conceived in the 1950s as a space bomber, which later became an orbit capable reusable spacecraft useful for various military missions in addition to the delivery of nuclear warheads. The spiritual successor to X-20 is the unmanned X-37, which if all goes well, will be launched into orbit in 2009 and later return for an unpiloted runway landing for later reuse. Essentially, the X-37 is a small unmanned space shuttle launched by expendable launch vehicle. The X-37 is being presented as the prototype for a future military space plane, able to reach orbit, perform the mission of a satellite, and return to earth on demand. In a way the autonomous space plane concept is a competitor to the concept of being able to upgrade and refuel satellites in orbit. Instead of autonomous on-orbit satellite rendezvous, servicing, and refueling, the military space plane would just land to be reused with the same or different payload. Like the space shuttle, various payloads may be mounted in the X-37 and its derivatives, including potentially maneuvering reentry vehicles such as being investigated under FALCON's secretive X-41 common aero vehicle (CAV) program.

Military space planes or other ground-based rapid reaction satellite replacement launch systems would be less vulnerable to attack than spares prepositioned in orbit. Orbit-based spares would be as vulnerable to attack as the satellites they are acting as understudy for. Ground-based satellite replacements could be stored within the protection of a military base. The concept of the *wooden round* or *all-up-round* (AUR) where munitions such as missiles are stored till needed in sealed containers with little maintenance could be applied to both SLVs and replacement satellites. Computer technology allows for the health and readiness of stored satellites and launchers to be checked autonomously, without exposing either to the outside world. Basing options for rapid reaction satellite launchers could expand to mirror the nuclear triad of ground, air, and sea basing of nuclear deterrent forces to ensure the survival of the U.S. ability to exploit space. Already orbital launch has been demonstrated from all three environments, and all that remains is demonstration of the desired reaction time. The capability for U.S. space power to be quickly rebuilt as part of a swift and devastating retaliation would perhaps be deterrence against an ASAT attack in the first place.

It is of note that warship design has largely abandoned armor in the balance between mission systems and everything else aboard. With missiles being a greater threat than large caliber shells, antiair defenses have arguably shown more utility than armor plate. Moreover, advances in microelectronics have allowed the scaling down of defensive antimissile/warhead systems to allow for active schemes to be mounted on tanks and armored fighting vehicles. Perhaps it would be more worthwhile to forgo the addition of shielding on satellites for other means of defense against attack. Small satellites in addition to being able to take on more space power roles from traditional large satellites can also be produced to provide protection for larger satellites. Small disposable satellites can be equipped with equipment meant to fool sensors, making them useful as decoys for larger mission satellites. Small satellites could also take a more active role in protecting another satellite. Doctrinally, satellite defense plans could come to resemble naval task groups by surrounding critical satellites with dedicated defense or bodyguard satellites ready to block or intercept attacks. This could potentially expand the defensive zone for a layered counter-counterspace capability from just the immediate area around an important satellite to be all of LEO space with force application technologies controlling just what is allowed into space, as discussed in the next chapter.

Electronic Warfare

U.S. space control policy has for some time desired a range of options for dealing with hostile and otherwise troublesome satellites. Electronic warfare (EW) techniques, which allow for temporary degradation or even denial of services from a target satellite, exist in a conceptual grey area between passive and active counterspace techniques. While outright destruction will provide a degree of certainty that a satellite will no longer be a threat, destruction often has operational and political consequences. While the subject of destructive space weapons produces headlines and vocal debate, electronic warfare methods of space control have garnered much less attention and debate. Aside from the fact that EW is among the most secretive of military arts, EW, even when in the open and labeled explicitly *offensive counterspace*, does not draw much outrage from opponents of space weaponization. Despite its seemingly inoffensive nature, using EW techniques to sever communications renders a satellite just as dead to warfighters as if it were shattered into millions of pieces.

EW is for many regarded as a black art of mysterious boxes that have an effect on the battlefield out of all proportion to their utilitarian appearances. Satellite communications jamming is, however, in many respects a proven technology. Essentially, radio frequency jamming seeks to drown out a signal with another signal on the same frequency. The addition of an unwanted transmission makes communications between satellite and ground station unintelligible severing the flow of information. The power requirements for interference are broadly similar to that of the transponder being jammed, and are otherwise unremarkable. The sheer simplicity of satellite communications jamming was demonstrated in the late 1990s by the exploits of Palapa B1.

Palapa B1 was an Indonesian-owned commercial communication satellite which was at the center of several disputes over the use of a geostationary orbital slot claimed by the island nation of Tonga and leased to third-party satellite operators. The dispute had the Indonesian satellite and competing satellites flown somewhat hazardously within the confines of the same orbital slot. This first occurred in 1992 when a Russian Gorizont commercial communications satellite, then being leased by American company Rimsat, was moved into the disputed orbital slot.[12] This incident led to discussions, which allowed Palapa B1 to continue use of the orbital slot. In 1996, Hong Kong–based APT Satellite Company was paying Tonga for use of the disputed orbital slot for another communications satellite, APSTAR-1A. Unlike the dispute involving Rimsat, the 1996 incident escalated, eventually culminating with Palapa B1 being set to transmit such that it interfered with APSTAR-1A, effectively jamming its communications with subscribers in the People's Republic of China till APT Satellite Company was able to successfully troubleshoot the interference. Both Palapa B1 and APTAR-1A are communication satellites, carrying powerful transponders capable of bridging the great distance between and GEO orbital altitudes. While the actions of Palapa B1 are not considered warfare, the ease with which it was able to interfere with APSTAR-1A clearly demonstrates the feasibility of the belligerent use of satellite communications.

The U.S. Counter Communication System (CCS), otherwise known as Counter-Com, is a purpose built satellite communications jamming system. CounterCom is a ground-based system that has been in service long enough with the U.S. Air Force for an expected upgraded Block 20 version. This is a modest system, with a budget only measured in the tens of millions of dollars.[13] The mobile CounterCom system is physically similar to portable military satellite communications systems. Conceptually the role is the same—to transmit a signal powerful enough to reach a communication satellite in GEO, except that for CounterCom reception is not quite as important.

Other examples of ground based satellite jamming are the GPS jamming equipment deployed by Iraq in 2003 to disrupt U.S. forces seeking to topple the regime of Saddam Hussein. The attempts by Saddam Hussein's regime to jam GPS were not very successful, as GPS-guided JDAM bombs were prominently used. In U.S. military press releases, it was noted that six jamming sites were discovered and were subsequently destroyed during the short air campaign prior to ground operations to remove the Hussein regime.[14] Russian companies are implicated in the supply of this technology. Other nations, notably China, are believed to be also working on local GPS jamming capabilities.

EW includes measures to overcome hostile electronic interference. Just as there are techniques to identify and interfere with critical communications, there are means to avoid interference. Modern military communications systems possess for the most part some degree of jam-resistance. Technologies involved with making communications resistant to jamming have found application in the domestic world. Digital spread spectrum (DSS) technology, a form of frequency jumping, is used in cordless telephones to reduce interference and improve ranges but had its roots in military counterjamming technology. Frequency jumping involves rapidly and seamlessly changing the radio frequency that is being used for communication, while keeping both ends synchronized. Military equivalents of DSS use a much larger range of frequencies and more complicated randomization schemes than that used for the household cordless phone.

Aside from creating difficultly in determining the correct frequency to jam, other countermeasures to interference and interception are also involved in robust military communications and for that matter in commercial radio frequency technology. Digital technology for instance includes error correction techniques—the addition of small pieces of extra information used to determine if what is being transmitted is valid. Packet switching networks, such as that found in the ordinary Ethernet computer network, not only break information down into fixed size packets, but embedded in each packet is information to determine where the packet is supposed to go and if the packet made it through intact.

Along the same direction to frequency hopping is move from radio frequency (RF) communication to optical communications. Lasers are already used to transmit a large fraction of the civilian world's data, via globe-spanning fiber-optic cables, forming the pipes of the information age. While most consumers depend on RF systems such as coaxial cable or plain old telephone-based high-speed Internet connections, the truly tech savvy consumer or business is willing to pay the premium for fiber-optic right to the house, relegating the cable/RF combination to backup status. Outside the glass fibers that bind the world's computer devices, laser signals are capable of impressive data rates as long as there is direct line of sight between both ends of the conversation. There are many challenges involved, in particular maintaining the tight beam on target so useful for preventing interference and dealing with the weather. The reason that GPS guidance is now in vogue with bomb guidance verses laser beam riding systems is that GPS is all weather. Of course, laser-based satellite communications are not a new concept. There was speculation during the final years of the cold war over the feasibility of a communication system based on lasers operating on blue-green wavelengths for relaying from satellites messages to submerged submarines.

One fundamental weakness of signals jamming is that it is easily detected. A signal powerful enough to drown out another powerful signal is easy enough to find with radio frequency direction finding techniques. The example of Iraqi GPS jamming ultimately resulted in the sources of the attempted interference being destroyed. It is noted with extreme irony that one of the sites was successfully attacked with JDAM bombs that require GPS update for their advertised level of precision. Arguably, the reason for Saddam Hussein's investment in GPS jamming was to mitigate the effectiveness of U.S. guided munitions in light of the well-publicized adoption of GPS guidance for weapons. GPS jamming does not take into account that systems such as JDAMs include a very accurate Internal Navigation System (INS), which in the case of JDAM's GPS backs up and not the other way around. Sharing the unclassified counterspace systems budget is the Rapid Identification Detection and Reporting System (RAIDRS)—a system that includes the capability to not only identify electronic interference, but also the location of the source. RAIDRS is the other side of the same coin as the CCS. The latter system interferes with satellite communications, and the former detects when such techniques are being used against the United States.

EW has always included that most prized of capabilities—the ability to insert false orders or information into the system. It is therefore only reasonable to expect that the ability to give a hostile satellite false orders is either an existing capability or highly sought after. On the other hand, breaking into the command and control loop of a satellite is not as easy as it would appear to be. Convincing a satellite to do something against the will of its owners on one level is easier than to give false orders to troops on earth. There is no psychology involved with satellites, as they are just machines, powerful in some senses but ultimately slaves to their controllers. Troops on the ground, on the other hand, may be able to determine that they are being led astray just by the tone of the false orders. However, satellite command and control requires highly specialized equipment and the ability not only to break the coded communications between satellite and ground station, but also to gain access to the actual command set for a particular satellite. It is relatively easier to jam satellite command and control signals to leave a satellite helpless and cut off from its controllers, than actually seize control of it.

Shutting down a satellite would not allow it to be turned against its owners in some intricate plan, but there are risks involved. If the intrusion is detected and the satellite's controllers play along until a critical moment, the controllers can effectively turn the satellite into a double agent. Satellite hacking is a less-straightforward military activity and resides more in the shady world of espionage. Spies are always after such secrets such as the ciphers that guard satellite communications. Even the secretive KH-11 satellites were compromised by foreign intelligence during the much more tense days of the cold war. In a global world where even military electronics are not necessarily made in the country of their use, there are fears related to satellite hacking, where back doors have been inserted into critical military systems. Countermeasures against this are part trade policy and part counterespionage.

A transmitter powerful enough to give commands to a satellite is powerful enough to be detected. This is where general satellite communications countermeasures play a role. Once a foreign control signal is identified, it can be jammed or attacked through terrestrial methods. Recent claims that a British Skynet military communications satellite was under threat of hijacking and even being held for ransom by supposed computer criminals (not known to be affiliated with any nation) are questionable due to the difficulty in secretly cobbling together the communications gear needed for just a caper. Then again, recent incidents by Russian hackers infiltrating the computer systems of Lithuania and Georgia during tensions between Russia and these two nations, and outright war in the later case, have certainly made clear that hackers do target military and national computer systems. In information-age and effects-based warfare, the work of talented freelancers or trained professionals may be indistinguishable.

Conclusion

Enhanced space surveillance and the emergence of passive space force application capabilities are relatively safe bets with regard to the future of U.S. military space power. Enhanced maneuver, shielding, and even deception with regard to satellite operations are more or less extensions of what is already being done today, only now with greater on emphasis on their use for security and defense. As nearly familiar space capabilities, passive counterspace has avoided offending the sensibilities of practically all except those who wish to drastically reorder the world to the detriment of the West. Even satellite communications jamming does not seem to cross the line with regard to opponents of space weaponization. That being said, these capabilities still demand research and development support to actually become operational.

Like all things dual-use, passive counterspace is a matter of intent. Specifically, passive counterspace capabilities are meant not to escalate things to open conflict. In earthly environments there are many outlets for competing nations to take things to the brink. Of late, Russian aircraft began probing Western airspace, reviving traditions of the cold war, but there is no serious talk of war as a result. Similarly, as long as the threshold for war remains high, passive counterspace measures would not be regarded as counterspace measures at all. While the orbital mechanics involved may be complicated, it is not totally inconceivable for a satellite today to be maneuvered so as to cause innocent signals interference. This was after all done before in a purely commercial dispute. The political costs and the costs to maneuvering capacity would have to be carefully weighed. However, it must be remembered that such traditions as aggressive close formation with foreign aircraft or naval vessels is often only the first step in politics by other means. Escalation from the fuzzy boundary of space weaponization to clear weaponization is not entirely out of the question, nor is it always undesirable. This leads to the more active forms of space force application.

Notes

1. U.S. Space Command, *Long Range Plan* (Colorado Springs, CO: USSPACECOM, 1997).
2. Peter Bond, *Jane's Space Recognition Guide* (London: Harper Collins, 2008), 234.
3. British Broadcasting Corporation, "Iran Launches Homegrown Satellite," February 3, 2009, http://news.bbc.co.uk/2/hi/middle_east/7866357.stm.

4. Paul Rincon, British Broadcasting Corporation, "Sat Collision Highlights Growing Threat," February 12, 2009, http://news.bbc.co.uk/2/hi/science/nature/7885750.stm.

5. United States Air Force, "Program Elements FY2009—Counterspace Systems," February 2008, http://www.js.pentagon.mil/descriptivesum/Y2009/AirForce/0604421F.pdf.

6. Dutch Space, "ConeXpress OLEV™," 2005, http://www.dutchspace.nl/uploadedFiles/Products_&_Services/ConeXpress/ConeXpress%20OLEV.pdf.

7. David S. F. Portree, "NASA, Mir Hardware Heritage," March 1995, http://ston.jsc.nasa.gov/collections/TRS/_techrep/RP1357.pdf.

8. United States Air Force, Air Force Research Laboratory, "Advanced Electro-Optical System," July 2002, http://www.kirtland.af.mil/shared/media/document/AFD-070404-028.pdf.

9. Leslie Hoffman, Associated Press, "Columbia Photo Taken with Simple Telescope and Computer," February 12, 2003, http://www.space.com/missionlaunches/sts107_starfire_030212.html.

10. National Aeronautics and Space Agency, National Space Science Data Center, "KH 11-10," http://nssdc.gsfc.nasa.gov/nmc/masterCatalog.do?sc=1990-019B.

11. National Radio Astronomy Observatory, "Welcome to the Very Large Array!" http://www.vla.nrao.edu/.

12. Eisenhower Institute, "A European Perspective on Current Trends in Military and Commercial Space," July 15, 2002, http://www.eisenhowerinstitute.org/events/past_events/old_events/071502MtngRpt.dot.

13. United States Air Force, "Program Elements FY2009—Counterspace Systems," February 2008, http://www.js.pentagon.mil/descriptivesum/Y2009/AirForce/0604421F.pdf.

14. Jim Garamone, American Forces Press Service, "CENTCOM Charts Operation Iraqi Freedom Progress," March 25, 2003, http://www.defenselink.mil/news/newsarticle.aspx?id=29230.

CHAPTER 4

Military Space and Force Application II: Active Force Application

The methods with which coercive force may be applied to, in, and from space are limited only by the laws of nature and imagination. Some systems are based on familiar and well-known weapons technologies used for years on earth. Weapons associated with science fiction are, however, more ingrained with the public when it comes to active space force application. Notwithstanding the fact that most science fiction breaks the laws of nature (as we know them at this time), it is not surprising that most if not all real space weapons concepts have equivalents in science fiction. Most works of science fiction, including the various iterations of *Star Trek, Doctor Who,* and even *2001: A Space Odyssey,* have a place for armaments and military organizations.

Active space force application in some sense is more easily defined than the passive means discussed in chapter 3. At the same time, there are considerations such as where a weapon platform is located relative to where its effect takes place, which raises questions as to how particular systems are classified as space weapons. Those distinctions, while largely the product of politics, will be touched on in discussing what is truly technologically possible in the weaponization debate. Indeed, it is the science and technology of what makes a space weapon that will ultimately define what can even be contemplated.

Historical ASAT Weapons

The method of space attack is limited only by physics and imagination (or paranoia). Anytime a spacecraft can be interacted with, it can be attacked. The emerging reality of space tourism raises the specter that subnational groups, in other words terrorists, may be able to strike U.S. assets in space. Most space technologies, even civilian ones, have potential for dual-use. Today, the world is deep within the fuzzy boundary between nonweapons types of space militarization and clear unambiguous space weaponization. This state has been reached with relatively little money and effort being spent on actual space weapons.

Arguably, an antisatellite weapon can be produced in a very short time span with off the shelf capabilities. Ad hoc weaponization could be as simple as ordering a satellite with enough maneuvering reserve to ram another satellite. The 1997 Mir docking accident proved just how much damage a low speed collision can cause. The vehicle at fault, a Progress supply craft, was designed for close proximity operations, specifically docking with space stations. At present, the European Space Agency has developed and put into space its own autonomous supply craft, the automated transfer vehicle (ATV), and Japan is preparing its own, the H-II transfer vehicle (HTV). Both ESA and Japanese spacecraft are meant to resupply the International Space Station; however, the technology they both embody are adaptable for weapons use. This weapons potential is only a side effect of these two space programs, neither the ESA nor Japan are presently known to have space weapons programs. The Soviet Union on the other hand did have its co-orbital ASAT program, which surely did leverage off of the Soviet space programs vast on-orbit rendezvous experience.

Paranoia over space projects can be seen largely as a product of its time. The space shuttle with its Canadian designed and built Shuttle Remote Manipulator System (SRMS), or Canadarm, certainly fed Soviet paranoia though the lens of the cold war. Soviet nightmares included the space shuttle pulling up next to a critical Soviet satellite, and menacing it with the Canadarm. The 1967 James Bond film *You Only Live Twice*, while not being very faithful to the novel, did suggest what a large cargo bay could do: hijack orbiting spacecraft. The space shuttle was, of course, advertised as having a large recovery capability, though operationally it only managed to salvage a handful of wayward satellites, and never at a profit. This was on top of fears that the space shuttle was a space bomber, using its large cargo bay not to contain captured satellites, but instead to house dozens of nuclear warheads. One has to wonder what the Soviet Union was planning to do with its own larger space shuttle, which was developed as a response to the potential of the U.S. program.

The ATV, HTV, Progress, and especially the U.S. and Soviet space shuttles are large spacecraft. The cost involved with each limited use, even for intended purposes. Cheaper small satellite programs have fuelled more realistic desires and concerns over space weapons. As mentioned in chapter 3, the United States is investing quite a bit into giving small satellites an autonomous capability to perform proximity operations. DARPA's Orbital Express program, which is aimed at extending the life of satellites though on-orbit servicing and refueling could be adapted easily to shortening the life of a satellite by having it rendezvous with a target satellite and then using its small robotic arm to wield all manner of weaponry against it. The U.S. Air Force's eXperimental Satellite System-11 (XSS-11) another microsatellite program has garnered more controversy due to its use in researching on-orbit satellite inspection, and claims that the basic XSS-11 satellite design is readily adaptable into an ASAT. Europe's ConeXpress OLEV is perhaps more threatening as a weapon. Its ability to latch onto an uncooperative satellite and propel it away from the orbit where it is useful represents the ability to attack a satellite and not leave a mess.

Compared to the KE ASAT and the fears of resulting space debris, the ability to make a satellite just leave the battlespace is more likely to be used. In reality all proximity operations, manned, unmanned, big and small are applicable for a range of uses from space station and satellite resupply, critical dockings needed for expeditions to the moon, and of course ASAT weapons.

At present there are no international agreements and conventions forbidding space weapons; hence there is research and development on actual ASAT technologies. The much-cited 1967 Outer Space Treaty (OST) does not ban space weapons in general. Article IV of the OST only specifically prohibits "nuclear weapons or any other kinds of weapons of mass destruction"[1] from placement in orbit and on celestial bodies. Article IV does forbid "the testing of any type of weapons and the conduct of military maneuvers on celestial bodies."[2] This, however, does not disallow the testing and conduct of military maneuvers at space altitudes or in orbit. Both the United States and Soviet Union during the years of the cold war conducted several antisatellite technology tests, which included the destruction of orbiting satellites. While there has been much recent discussion outside the United States over prohibiting space weapons, other nations have also been active in the area of space systems, which are undeniably space weapons.

On January 11, 2007 the People's Republic of China conducted its first successful ASAT test. The Chinese ASAT designated SC-19 in the United States, is a direct ascent interceptor, thought to be launched by a multistage solid fuelled rocket. SC-19's kill mechanism is a kinetic kill vehicle, though at the present time, the form of terminal guidance used is unknown. FY-1C a defunct, but still under control, Chinese weather satellite in polar orbit was the target of this test, and was impacted above mainland China near Xichang Space Center at an approximate altitude of 530 miles[3] (850 km). It is unknown if Xichang Space Center was the actual launch point as the presumed solid fuel launch vehicle may be based on Chinese ICBM technology, which includes missiles that can be launched from mobile transport erector launchers (TEL) trucks. China Aerospace Science and Industry Corporation (CASIC) market the KT-1 solid fueled launch vehicle, which forms the basis for many estimates of performance for the complete SC-19 ASAT weapon system. The People's Republic of China did not acknowledge the ASAT test till January 23.[4]

Some view the test as a paradigm shift in Chinese diplomacy favoring limitations on space weapons. Then again SC-19 fell outside the scope of Chinese space weapons diplomacy. The 2008 Chinese-Russian proposal for a space weapons ban, specifically addresses weapons placed in space, where *placed* is defined as orbiting, including fractional orbits.[5] The wording of the draft exempts ground-based offensive systems, such as SC-19, but does include defensive systems for the protection of satellites, so-called body guard satellites, and orbiting missile defense systems. Chinese efforts in the UN supported Prevention of an Arms Race in Outer Space (PAROS), similarly does not address SC-19. So at the very least China has been consistent with the letter of their diplomacy, which would leave loopholes for activities that it views as beneficial.

Surface-based missile defense is another existing capability that can be adapted into an ASAT. Missile defense divides the interception of a ballistic missile into three phases: boost, midcourse, and terminal. Midcourse in a ballistic missile trajectory is the stage after the missile's rocket engines shut down and the payload of warhead(s) and any included decoys coast through the vacuum of space on a purely ballistic course until terminal phase when the warhead(s) reenter the earth's atmosphere. For even a medium range ballistic missile, this altitude tops off well above the usual delimitations for space. Therefore midcourse missile defense interception occurs in space, giving argument to the case for earth-based midcourse missile defense capabilities to be defined as space weapons. The longer range the ballistic missile, the further away from earth the midcourse phase will take it and the more into space a midcourse missile defense system will need to venture to conduct its attack. As the U.S. fields an ever improving midcourse ballistic missile interception capability it raises the question of what else can be intercepted by missile defense. On February 21, 2008, this was answered with the interception of a wayward U.S. satellite by a SM-3 missile.

In contrast to the Chinese ASAT test of 2007 this U.S. interception in 2008 was against an uncooperative target, USA-193, a U.S. National Reconnaissance Office satellite that failed shortly after being orbited. This satellite had a full load of hydrazine monopropellant[6] onboard. Hydrazine remains very stable for a long time, which makes it useful for maneuvering thrusters found on satellites. However, hydrazine is toxic, and due to the quantity present on the large satellite there were concerns that a quantity of it could have survived reentry perhaps impacting in or near inhabited land as Skylab did in 1979. As USA-193's orbit was already faltering, the debris from its destruction deorbited within a few weeks to months as opposed to the years and decades for FY-1C. Smaller pieces, including the shattered hydrazine tank, would easily burn up on reentry, reducing any hazardous materials threat to at worse air pollution.

USA-193 was shot down by a modified U.S. Navy RIM-161 Standard SM-3 missile fired from the cruiser USS *Lake Erie*. The SM-3 is itself a development of the Standard family of naval surface-to-air missiles in use since the late 1960s by the United States and allies such as Canada and Japan. Instead of carrying an explosive warhead, the payload of the SM-3 is a kinetic kill vehicle (KKV) with an imaging long wave infrared sensor based terminal seeker. Prior to release of the KKV, the SM-3 receives guidance from the modified AEGIS weapon system onboard the launch ship upgraded for the ballistic missile defense mission. Additional sensor support for the interception was provided by other AEGIS equipped ships accompanying the *Lake Erie*, and other elements of U.S. aerospace tracking.

The fact that SM-3 is not a dedicated ASAT illustrates the similarities between the missile defense mission and the ASAT mission. The U.S. Navy's SM-3/up-rated AEGIS warship combination is designed to intercept short and medium range ballistic missiles during the target's midcourse portion of missile flight. It must be remembered that the successful SM-3 program is meant to intercept at lower altitudes than the U.S. Ground-based midcourse defense (GMD), a system meant to destroy long-range ballistic missiles. The much large GMD missile is not mobile in its present form and therefore cannot be as easily parked underneath the expected orbital path of a target.

Historically, both the United States and the Soviet Union have deployed what were unambiguously ASAT weapons. The Soviet co-orbital ASAT is reputed to have been operational, though of somewhat questionable utility. In the United States, Project 437 achieved operation status from 1964 to 1975, though it must be noted that a functional weapon system did not exist at all times during this period. Project 437 comprised of a Thor intermediate ballistic missile (IRBM) armed with a nuclear warhead set to go off near an orbiting (or fractionally orbiting) target. The Thor IRBM eventually spawned the Delta family of orbital launch vehicles. The nuclear warhead ASAT spawned a lot of questions over the consequences of its use.

The devastating, indiscriminate, and long-term ASAT effects of a nuclear weapon detonating at high altitude were demonstrated (accidentally) during the 1962 Starfish Prime high altitude nuclear test. A 1.4 megaton nuclear detonation at a mere 400 kilometers altitude was able to generate radiation belts (similar to the natural Van Allen belts), which over time disabled several satellites before naturally dissipating. Artificially elevated radiation conditions can last for months or even years depending on the yield and number of warheads detonated. During this time frame, a satellite could easily absorb more radiation that it was meant to in its designed life span and fail from excess radiation exposure.

A nuclear event in space would have other more immediate effects: these include radiation damage to satellites caught within line of sight, and nonline of sight damage such as the EM interference experienced in Hawaii (roughly 1,500 km away from the Johnson Island test sight) during the Starfish Prime Test. Unlike the kinetic kill system currently favored now by the United States for near-term missile defense or China's 2007 ASAT test, a nuclear kill mechanism could get away with a 'near-miss' of several dozen kilometers. A nuclear ASAT would not require the complex sensors or precision guidance of nonnuclear U.S., Chinese and Soviet systems. The brute effectiveness of a nuclear warhead was also used during the Cold war for ballistic missile interception in systems such as the U.S. Nike Zeus[7] and the Soviet Galosh missile batteries, which defended Moscow into the 1990s.

The Nuclear ASAT is essentially a nuclear warhead on a rocket capable of reaching the orbital altitude of prospective targets; therefore every one of today's nuclear armed states have the potential for this type of ad hoc but supremely effective means of space denial. The U.S. Army (After Next) 1997 Winter War Game gave a taste of what would happen to U.S. space power if faced with a near-peer that felt it had nothing to lose. In the scenario presented in the 1997 War Game, the world of 2020 would pit the BLUE Team (a future U.S. military based on transformational concepts) against the RED Team (a modernized Russian military but not as transformed as BLUE Team) in an European conflict that fell outside of the threshold for all out nuclear war. The RED Team, expected to only give a convincing but ineffective fight, was able to cause some early upsets, including the very effective use of nuclear ASAT weapons. Without the intervention of the judges to override the devastating damage to BLUE Team's space infrastructure, RED Team's early use of ASAT weapons effectively produced an early victory. The RAND Corporation's study of the 1997 Winter War Game is freely available to all with Internet connections.[8]

Outside of games, the use of nuclear weapons between nuclear-armed great powers, even for tactical purposes, risks uncontrollable escalation. The end result of runaway escalation being mutually assured destruction (MAD) for all participants and quite likely the end of civilization as we know it. Short of all out nuclear war, the indiscriminate nature of nuclear weapons also make them less appealing to technologically sophisticated nations as they would jeopardize their own investments in space. Unlike the RED team in the 1997 Winter War, the real world near-peer would have to deal with the aftermath of nuclear space denial.

The nuclear ASAT has appeal beyond that of the established nuclear powers that participate in the global economy. Indeed making the world a poorer place would be a great equalizer for some on the international stage. U.S. space superiority faces threats not only from technologically sophisticated near-peers, but also from rogue nations who may become successful in acquiring nuclear weapons and the means to deliver them. A rogue nation, unlike a near-peer, is arguably less inclined to follow the norms of nuclear deterrence. The rogue nation using nuclear ASATs to blind and disorientate the United States, indeed the world, may be the only contemporary scenario where a nation could get away with the tactical use of nuclear weapons. This creates the specter of the nuclear-armed rogue state able to deny the United States and its allies their space force enhancement capabilities via a nuclear ASAT weapon.

For the price of a few nuclear-armed sounding rockets and international condemnation, a nation can take out earth imaging (spy) satellites owned by the United States and others. Working beyond basic nuclear weaponization to produce smaller warheads and more powerful rocketry, the emerging nuclear weapons state can put the Global Positioning System (GPS) constellation at risk. If the rogue state is allowed to develop the ability to kick a payload out to the distances of a geostationary orbit, satellite communications, the most vital of transformational military relevant space services, is jeopardized. It must be noted that for both the United States and the Soviet Union, the ability to place a satellite into a proper geostationary orbits was achieved in the 1960s; a nuclear ASAT does not need that kind of precision. The nuclear ASAT gives a nation the ability to turn off the critical space lines of information. Loss of a major space system would be costly to the advanced nations of the world, possibly crippling. For the rogue regime these are only luxuries (compared to basic survival); for the people caught under the thumb of these regimes, space is inconsequential.

Without support from space, technologically sophisticated militaries are rendered relatively powerless. The rapid and relatively low cost (in terms of force investment and casualties) collapse of Iraq's conventional military in 2003 was dependent on space systems. Cut off support from space, the small high-tech transformational force capable of subduing larger less advanced opponents becomes just a small military force, overburdened with expensive but now useless equipment. While there has been a lot of overselling of military transformation in recent years, nobody in the West wants to go back to the old ways of doing things, especially the acceptance of high number of casualties for both sides in a conflict. On the other hand those opposed

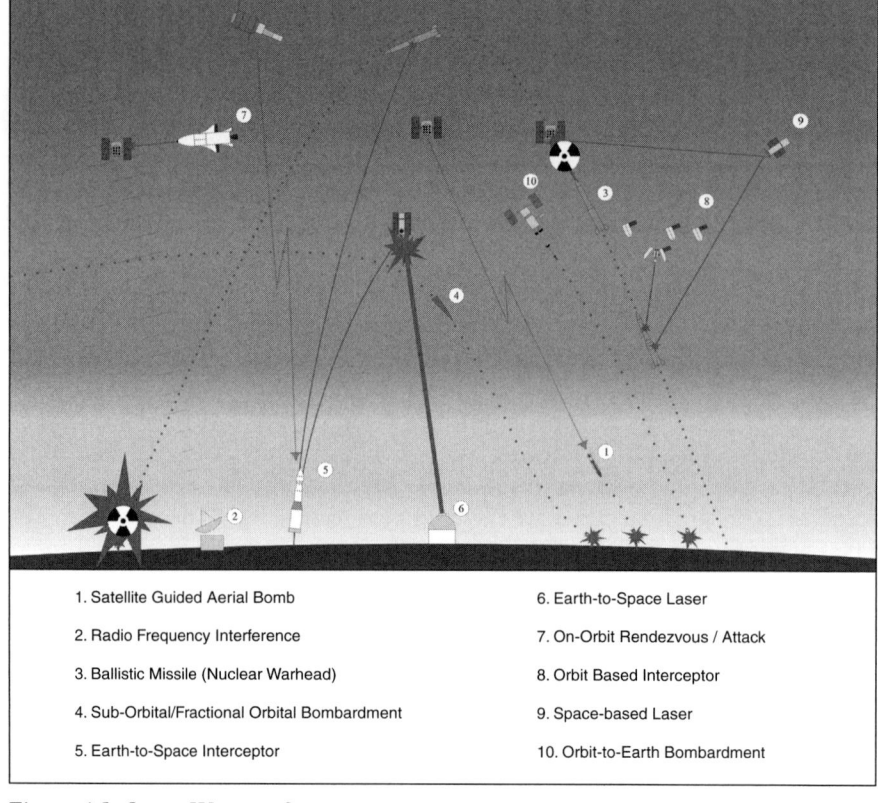

Figure 4.1 Space Weapons?

to U.S. and Western policy understand that the United States and the West is casualty averse. For these pariah nations, casualties and destruction are part of how they do business.

This usage of nuclear weapons in the ASAT role against U.S. space power echo's another era's use of nuclear weapons to offset another era's dominant conventional military paradigm. Not long ago the massive numerical superiority of Soviet armored forces (the conventional warfare paradigm of its day) was countered by NATO's tactical nuclear weapons. It was only later in the cold war that precision guided conventional munitions offered a nonnuclear means to counter the Soviet numerical advantage. Many of the systems vital to the current state of the art in warfare are space based and therefore under threat from a nuclear ASAT. There is great irony that the precision warfare systems that displaced nuclear weapons as the decisive weapon on the battlefield are now being put under threat by another tactical oriented nuclear weapons usage. The nuclear ASAT is in this context the Tac Nuke of the information warfare era.

The tactical nuclear weapon parallel with a nuclear ASAT is apt in that tactical nuclear weapons were meant to be used in a limited fashion, confined to the battlefield and to a large extent sparing cities and populaces. Many generations of scholars and strategists have argued over the validity of the concept of tactical nuclear weapons being confined to tactical targets. One major concern is that once nuclear weapons are used against fielded military forces there is nothing to stop retaliation from escalating to use of nuclear weapons against other strategic weapons (counterforce) and then to cities (countervalue). This scenario is based on a conflict involving nuclear-armed superpowers. The impoverished rouge nation that manages to obtain and use nuclear weapons in an ASAT capacity is a different scenario. Mutually assured destruction is an unlikely outcome as the rogue nation is unlikely to have many, if any, long-range missiles capable of threatening the United States. Moreover, the decision to use their limited nuclear capacity in an ASAT role targets largely unmanned U.S. space assets. There are many, on all sides of U.S. political life, that find troubling the concept of nuclear deterrence and the avenging of what Herman Kahn coined as megadeath with more megadeath. The idea of avenging several billion dollars of hardware and the ability to wage high-tech war is even more dilemma inducing that that of straightforward nuclear deterrence. However, a nuclear ASAT attack may deny the United States the ability to intervene regionally, allowing the rogue nuclear armed state a victory in its regional war and to effectively get away with using a tactical nuclear weapon.

Kinetic Energy

There is nothing more certain of neutralizing a target than to smash it into millions of little pieces, the basic concept of the kinetic energy (KE) weapon. Kinetic energy is the energy of an object in motion and, as everything in orbit is in motion at an extremely high velocity, any impact potentially involves a devastating amount of KE. While this is a surefire way of achieving destruction in space, the hypervelocity-attack concept does not lend well to dialing back effects. Clearly this is more of an old-fashion type of attack, with none of the refinement desired by proponents of military transformation. Moreover the results of a KE attack would leave behind orbiting debris, which would put at risk the attacker's own satellites. Despite its drawbacks, the KE attack is a nation's earliest counterspace capabilities.

$$KE = \frac{1}{2} (Mass) (Velocity^2)$$

Kinetic energy increases by the square of the velocity involved. Doubling the mass of the KE weapon (KEW) would only double the energy of the collision system. Doubling the velocity of the KEW would quadruple the energy of the collision system. This is the attraction of KE in a space weapon; negligible air resistance allows objects to reach very high speeds in space. Putting a relatively stationary object in the path of a satellite means a closing velocity of at least the Mach 25 speed needed for the target satellite to stay in orbit. The energy released by a few dozen kilograms of

KEW would shatter most satellites. The energy released by a few hundred kilograms of KEW at almost orbital speeds impacting the earth would involve more energy than the weapon's weight in high explosives. Then again the chemical energy needed to orbit the KEW or target satellite in the first place is substantial. In this respect the kinetic energy of the system is the result of a launch vehicle converting chemical energy into motion. Essentially a space KEW involves energy of a space launch, in the case of an ASAT the much larger energy needed to launch the target satellite.

Pundits often claim that once a nation achieves long-range missile technology it also achieves a basic direct-ascent ASAT capability. Direct-ascent interception refers to the weapon rising from the earth to intercept a target. A ballistic missile does not achieve orbital speeds, but instead can reach altitudes of thousands of kilometers, putting within reach LEO and even some MEO satellites. Littering the space in the immediate path of an orbiting satellite with debris or pellets, as one would use caltrops against horse cavalry and wheeled vehicles, is often suggested as an example of how simple it would be to destroy a satellite. The ballistic trajectory of the debris or pellets would not provide much kinetic energy against the satellite, indeed if the attack were to occur near the apex of the ballistic trajectory the speed would be at near zero, as gravity has either countered all of the upward velocity imparted by the rocket, or has already begun to pull it back to earth. Instead the target satellite's orbital speed would provide the energy to shatter itself, as all this simple kinetic energy attack has to do is put something reasonably massive in a position to be rammed by the target satellite.

The uncertainty of the true position of any satellite, at any given time, is dealt with by releasing a cloud of debris. The major impediment to this type of attack is the requirement of getting the cloud of debris into the path of a satellite so that the convergence of the two trajectories results in a destructive impact. Though the fine degree of missile flight control shares much with ICBM development, the space tracking needed to locate a target satellite and calculate its motion with exacting detail is hard to obtain, unless the target is a commercial satellite whose orbit is in the public domain. This can be overcome with wider areas of attack (putting more debris in space), but as the low incidence of satellites being disabled by natural and artificial space debris demonstrates, the chances of such a primitive ASAT successfully hitting its target before its ballistic course brings it back to earth are marginal.

A direct ascent flight profile can also put a hit-to-kill kinetic energy warhead into position to attack. A hit-to-kill warhead, or more descriptively, a hit-to-kill vehicle, contains sensors, propulsion, and enough computer power to recognize the target and order maneuvers to ensure a collision course. The hit-to-kill attack does not depend on luck, or even a buckshot effect, but instead information age technology such as solid-state microelectronics and sensors. It also helps that the entire payload hits the target in the hit-to-kill warhead, maximizing the destructive effect. This is the preferred method of in space KE attack for technologically advanced nations of the world. The January 2007 Chinese ASAT test is thought to have been a direct ascent ASAT. The U.S. Air Force's F-15 fighter launched Air Launched Miniature Homing Vehicle (ALMHV) demonstrated U.S. potential for fielding a direct ascent

KE ASAT in the 1980s. Several fielded ballistic missile defense systems, such as the U.S. Navy's Standard SM-3 and the longer range Ground-based midcourse defense (GMD) use hit-to-kill warheads, though these depend on the much lower, but still respectable, kinetic energy of two missiles set on collision course.

Going in the other direction are orbit-based kinetic energy weapons. Both the orbit-based SDI Brilliant Pebbles concept and the earlier Space-Based Kinetic-Kill Vehicle (SBKKV) strategic missile defense concept were based around hit-to-kill KE missiles equipped with powerful rocket propulsion to perform an intercept from orbital basing. The SBKKV concept had multiple interceptor missiles housed inside large orbiting 'garage' satellites that protected the missiles from the space environment and provided command and control for the missiles after launch, essentially an orbiting missile battery. The guided missiles of SBKKV were referred to as smart rocks, an allusion to the smart bombs and missiles then gaining prominence in the mid-1980s. The relatively large SBKKV garage satellites not only made deployment expensive, but also made for tempting and arguably easy targets for Soviet ASAT weapons. *Brilliant Pebbles* was based on the expectation that American technology would soon be able to produce a smaller, yet more capable missile, each responsible for its own autonomous orbital operation, including the decision to attack, hence a brilliant pebble. Several thousand independent Brilliant Pebbles interceptors would be maintained in orbit, which reduced vulnerability of the system to ASAT attack and eased the problem of conducting an intercept. With thousands of Brilliant Pebbles interceptors, each satellite only had to watch and if necessary carry out an interception over a smaller piece of the global battlespace. The Brilliant Pebbles concept treated satellites literally as munitions; no matter how brilliant it was still built to be expendable and hence affordable even in numbers required. Years after the end of the cold war, the technology needed for producing Brilliant Pebbles has only advanced, especially in the area of computers. Despite the refocusing of U.S. missile defense efforts on terrestrial basing and targeting against limited goals, the Brilliant Pebbles concept is still suggested when effective boost-phase missile defense options are considered.[9]

A more advanced form of motive force for KE warheads that should not be omitted is the electromagnetic railgun. A powerful electric current is run from one rail to another through a conductive projectile, generating a powerful Lorentz force, which accelerates the projectile along the two rails. Rail guns are capable of much higher acceleration than chemical-reaction-based gun propellants. The extreme 'muzzle' velocity of railguns has this technology examined for use in all environments, including examination during the early phase of the Regan-era SDI as a way of launching guided KE interceptors from a smaller number (relative to Brilliant Pebbles) of orbital platforms. Like directed energy weapons, power requirements limited near-term exploitation of railguns for orbital use, though the technology is being considered for use in terrestrial artillery applications such as naval bombardment, tank main armament, and land-based indirect fire. Magnetic launch is also proposed for cheap space launch of robust payloads that can withstand the many gravities of acceleration produced by a railgun such as electronics and fuel supplies.

Power from terrestrial sources is easier to obtain, especially for the fixed space launching railgun.

The Soviet Union was not left out in field of destructive kinetic energy space weapons. Soviet fighter cannon armament was mounted on a few military space stations of the Almaz series.[10] The Soviet surface to space co-orbital ASAT, which was tested and perhaps even became operational briefly during the cold war was both guided and used a kinetic effect as its kill mechanism. Described as a hot-metal kill weapon, the co-orbital ASAT once brought within attack range by its guidance system would explode its high explosive warhead to send fragments toward its intended target. Lack of atmosphere means that no concussive effects will result. This leaves the destructive energy of the high explosive to be conveyed by the hot-metal fragments. In general, it should be noted that familiar earthly bullet shooters and fragmentation bombs may also be called a KEW,[11] though usually this is reserved for projectiles that are moving at significant speed on impact.

There is no question as to the effectiveness of a kinetic energy weapon against satellites; however, the aftermath of a KE attack is a major impediment to its use. An energetic satellite breakup, such as that resulting from a KE attack, will result in untold numbers of fragments. The debris cloud from a shattered satellite will spread out from the trajectory of the original satellite. An orbiting debris cloud will continue to orbit until atmospheric resistance brings it down to earth. The higher the altitude the attack takes place, the more feeble the already thin atmosphere will be. The longer debris stays in space, the greater the chance it will become involved in an unintentional impact. Essentially a successful KE attack will result in the target turning into a long duration mass debris attack of its own. The Chinese MEO ASAT test of 2007 resulted in a debris cloud that is estimated to remain an orbital hazard for years. The ultimate fear is the cascading generation of on-orbit collisions of the Kessler syndrome mentioned earlier.

Space debris knows no loyalty, only the path given to it by physics. The United States, being the most satellite dependent nation on earth, is therefore most at risk from space debris, no matter how it is generated. At the very least it would be embarrassing, to have a satellite important to U.S. national security destroyed by debris left over from a kinetic energy ASAT attack conducted by the United States. It is for this reason that in many space security circles, there is not much enthusiasm for the KE ASAT. It must be remembered that the threat of space debris does not exist for ballistic defense applications of kinetic energy. The targeted ballistic missile does not have a trajectory, which will result in debris orbiting. Orbit-based antimissile systems generally have trajectories, which will take it and its remains directly into the earth's atmosphere. U.S. space control policy seeks to make space safe for U.S. use, and therefore methods of attack, which make space a more hazardous environment are of questionable utility.

Directed Energy Weapons

Directed energy weapon (DEW) technology commonly springs to mind for the subject of space weapons. The ability to project destructive energy without physical projectiles, the clichéd death ray is often used to define a superior technology to science-fiction audiences and policy makers alike. Unlike brute force legacy techniques such as bullets or bombs, DEW systems provide options to control how much destruction can be visited on a target. The potential for directed energy weapons to have precision in both targeting and in tailoring effects fits well with the ideals of the current information age mode of warfare employed by the United States. Other characteristics of directed energy technology promise unprecedented range, engagement speed, and simplified logistics, as long as the technology is supported to maturity. For some directed energy weapons form the core of next transformation in military affairs.

A directed energy weapon is exactly what the acronym spells out, a weapon based on energy being transmitted in a specific direction. In the vacuum of space, DEW technology means the projection of electromagnetic (EM) energy. While in nature, energy has a tendency to radiate in all directions; human applications have sought to project energy in specified directions. Both a burning torch and a laser are examples of harnessed EM radiation; fire is perhaps the first, the latter among the most recent. The lit torch provides illumination in all directions, the laser in only a very narrow beam, which in many applications is the very definition of direction.

Energy when deposited on a target can cause different effects based on the type of energy being used and what is being targeted. EM radiation can setup temporary electrical effects, which interfere with satellite operations, damaging electrical effects, destroying electronics, or rapidly heating materials and components to well beyond their failure point. Deliver enough energy quickly and the targeted material will rapidly convert to vapor, effectively exploding. Such fine control over the intensity of a DEW attack fits in with the desires of U.S. policy makers for counterspace options ranging from temporary effects to outright destruction.

A directed energy weapon will fire as long as it is supplied with an energy source; depending on the energy source used this could be potentially a limitless magazine. EM energy generally moves much faster than physical means of attack, at or near the speed of light, approximately 1.1 billion kilometers per hour (299,792,458 meters per second in a vacuum).[12] For practical purposes, an EM-based DEW will hit its target the instant it is triggered even at ranges measured in hundreds or thousands of kilometers. The speed of attack reduces the timeline of an attack to just that needed for the weapon's kill mechanism to operate. Rapid engagement times plus the potential for practically unlimited firings mean that a DEW platform will be difficult to overwhelm.

Revolutionary concepts, real, potential and imagined, always garner opposition, and DEW technology is no different. Chemical energy systems, the low explosives at the heart of most firearms, and the high explosives at the heart of most conventional warheads, represent a compact, effective and very familiar source of destruction. Powering directed energy weapons remains a major stumbling block to acceptance. Large quantities of electrical or other energy must be harnessed over a very short time span for a destructive DEW to function. Ironically, explosive driven mechanical-to-electrical power conversion technologies are proposed as possible solutions to the DEW power problem, though single-use cartridges filled with high explosives and high-tech electronics does not lend well to the *limitless-magazine* concept touted by DEW supporters. Related to the energy supply problem for DEW concepts is the waste heat problem. Energy conversion from one form to another, electricity to light, for example, is an inefficient process, with heat as one of the usual by-products. Even with improvements in efficiency the large quantities of energy involved with a destructive DEW attack will still mean a large amount of waste heat is generated. While the destructive capabilities of DEW are debatable, there is no question as to the destructive effects of excess heat on a spacecraft such as a DEW platform. The fielding of DEW requires that these many, often unprecedented, technical challenges be overcome. Ultimately for some DEW represents an expensive and redundant method of achieving destructive and even tailored precision effects.

Today in the early 21st century, directed energy weapons are no longer purely science fiction. Leading the pack for emerging DEW systems are relatively low-power DEW systems such as the U.S. active denial system (ADS). Indeed the ADS, a less-than-lethal crowd control system, has run into some controversy, not over ineffective power but instead over whether the energy levels delivered are too high for a less-than-lethal weapon. ADS operates on the millimeter wave part of the EM spectrum, more associated with relatively safe radar and RF communications than the microwave-based cooking appliance found in most American kitchens. Criticisms over how ADS is employed aside, it has on a limited scale demonstrated much of the promise of direct energy weapons. The mobile ADS can project its effect, a sensation of extreme heat, in a wide arc or in a targeted beam fulfilling the desires of adaptability and precision. The ADS is effective beyond the range of small arms, and the effect of an "intolerable heating sensation"[13] on its target is rapid. Powering ADS does not require anything too exotic. It can be powered by mobile generators. A fixed ADS emplacement plugged into the local power grid has an effectively limitless magazine. The technology has matured enough for Raytheon, the manufacturer of ADS, to offer a scaled down version to law enforcement and other domestic agencies for crowd control and infrastructure protection applications.[14]

Moving up from the pain beam of ADS are more destructive and clearly military oriented high power radio frequency (HPRF) weapons. As the intended effects include destruction the technologies being investigated include high power microwave (HPM) systems. Instead of drowning out a legible signal with RF noise or tricking it into accepting a false one, a high power RF or microwave attack directly corrupts information within electronics and computers by interacting with the materials of a circuit setting up electrical and magnetic effects. This requires hitting a target with enough energy not only to setup the desired EM effects but also to overpower any protection that may be present. Surge protectors are not perfect, not in the office, and not in military applications. At increased power levels those electrical effects will permanently destroy electronics through sparking and electrical arcing, though not as spectacularly as shown in films such as 1995's *Goldeneye*. The *Goldeneye* weapon that fictional MI-6 agent James Bond was out to stop was an orbiting nuclear bomb powered electromagnetic pulse (EMP) weapon. The movie further alludes to EM warfare with a cameo by the Eurocopter Tiger helicopter prototype, reputed in the script as being EM proofed.

The real world aviation industry is well aware of the dangers of EM interference. Allegedly, several military aircraft have crashed due to accidental interference caused by flying too near high power civilian broadcast transmitters. Many modern warplanes, likely including the real Eurocopter Tiger, now in service with several European nations and Australia, are hardened to some degree against the EM effects of a nuclear battlefield. A more ubiquitous example of aerospace concerns over EM interference is the requests for passengers to turn off mobile electronics, in particular cellular phones when airliners are taking off or landing. Modern airliners are dependent on the function of electronics and computer systems to stay safely airborne. Not all electronics work together, leading to the great expense involved with system integration of electronics from different vendors. Standards for EM interference and shielding against interference cannot be expected to be infallible. The sheer number of consumer electronics available around the world, each a potential source for EM interference, would be cost prohibitive to test against. Moreover if only a handful of consumer devices caused problems, there would be the problem of specifically prohibiting these devices, better to have a blanket prohibition against all consumer electronics. Now it should be remembered that aerospace and consumer electronics are not meant to produce damaging signals.

There is potential to scale the effects of a HPRF weapon. Control over power and frequency potentially allows a HPRF simply to jam satellite communications. Increasing the power level, the effects could range from temporary satellite malfunction or permanent crippling but with reduced possibility of the satellite breaking apart, thus avoiding the problem of debris. Moreover, temporary malfunction or permanent electronics failure could be attributed as satellite failure due to natural causes if the target satellite operators are not equipped to detect such an attack. Most satellites and the agencies that operate them are not well known for real time attack monitoring, though satellite post mortems are regularly conducted. Still, the subtle nature of a limited but damaging RF attack opens the opportunity for covert attack

On the other hand, the lack of visible effect from a HPRF attack leads to the problem of determining the success of an attack. The effectiveness of ADS as a crowd-control DEW on earth is known simply by observing the targeted individual or individuals fleeing from the pain beam. A satellite degraded or even disabled by a DEW attack would not necessarily show any damage or other signs that an attack had occurred. More dangerous than ineffectiveness, the aftermath of an attack could be simulated by a satellite's controllers playing possum to lull the attacker into a false sense of security. While near term research and development puts within reach the option of a stun setting, technology cannot yet peer into the black boxes of targeted equipment to determine just how 'stunned' they really are. However, like all DEW technologies, HPRF is a power hungry application. While Raytheon's ADS is a first step for the deployment of this technology, it must be remember it is only a comparatively short range, low power less-than-lethal weapon.

Among the proposed antimaterial HPRF weapons are less sophisticated single-shot weapons. In addition to harnessing the EMP of a nuclear weapon, there are concepts based around the conversion of the sharp sudden release of energy from the detonation of a high explosive into an intense pulse of electrical energy.[15] Reusable destructive HPRF weapons are being investigated for naval applications where power would be available from the great electrical generating capability of a future warship integrated with full electric propulsion. Using the engines of a warship to power a weapon would satisfy the limitless magazine potential, as long as the warship had fuel. RF and microwave-based DEW systems are of particular interest for use in the maritime environment due to the fact this type of transmitted energy is not too badly affected by the atmospheric humidity. Space applications face a situation opposite to those found by navies; power capacity is limited, but there is no atmosphere to attenuate the beam.

Another technology with great potential to project a tight beam across great ranges, and perhaps most greatly association with space warfare, is the laser. Light amplification by stimulate emission of radiation is the method for generating an intense and coherent beam of light and the origin for the word laser. Laser light is coherent; essentially this means that unlike natural light, which goes in all directions, a laser will produce a beam of light with practically no divergence from a single direction. The tight focus of a laser can reach out to a significant distance. Laser-based sensors have proven quite useful for obtaining precise observations of space objects from earth, including several capable of bouncing a light off reflectors left behind on the moon during the U.S. Apollo missions. Since the Vietnam War, the U.S. military has exploited the ability of a laser to paint light onto a target at distances of several kilometers. Currently, laser designators only guide conventional munitions to targets. A laser-based DEW would deposit enough energy to be the destructive mechanism itself.

Just as there are different types of light bulbs, there are many ways of generating laser light. The first laser was based on a ruby crystal lasing (or gain) medium, placed in between two mirrors, which formed an optical cavity, with one mirror partially transparent to allow the generated beam of laser light to be emitted. An energy source stimulates the lasing medium into emitting coherent light. In the first laser an external flash lamp powered or pumped the lasing medium. The first laser beam was generated in 1957 at Bell Labs. It is of note that the microwave amplification by stimulated emission of radiation (MASER), which relies on similar physics, was demonstrated earlier in 1953 by the same Dr. Charles Townes credited with inventing the laser. The nomenclature of laser does not just cover a device operating on visible light wavelengths. Hence the MASER can also be called a microwave LASER. Similarly, a laser operating on gamma-ray parts of the spectrum are sometimes referred to as a GASER or GRASER.

Energy for pumping a lasing medium can come from direct electrical discharge, noncoherent light (as in the original laser), another laser, a chemical reaction producing a both lasing medium and light, or nuclear reactions. The purpose of a laser to a large extent determines the combination of energy source and lasing material. Small efficient solid-state diode lasers are found in domestic low power applications ranging from invisible infrared laser light supermarket scanner to visible light laser pointers. Carbon dioxide and other gases can be pumped to produce lasers useful for industrial and medical applications, close range cutting and burning. Chemicals reacting inside an optical cavity both provide energy and act as the lasing medium can deliver energy levels measured in megawatts and hence form the basis for near term weaponized lasers. The nuclear weapon pumped X-ray laser studied during the Reagan-era SDI program emitted powerful beams of coherent X-rays. Clearly, there is plenty of power released by a nuclear detonation; however, this energy source would run afoul of many political obstacles, in addition to the already difficult problem of generating a laser beam in the very brief time span of a nuclear event.

The airborne laser (ABL) now in testing, the tactical high-energy laser (THEL), and the proposed space-based laser (SBL) are all based around powerful chemical laser reactions. The nature of the chemical reactants, which power these lasers, defines the output laser beam. ABL and THEL both operate within the earth's atmosphere, and hence the reactants used in both were selected in part due to the need to produce light wavelengths that are not rapidly absorbed by the atmosphere. Work on the short range THEL system is now completed, and the technology is being marketed as a potential defense against mortars, rockets and artillery shells. ABL, while behind schedule, is still making progress toward the eventual goal of demonstrating the ability to destroy short and medium range ballistic missiles at hundreds of kilometers of range. The SBL chemical laser, though far from the deployment stage, is envisioned to have an effective attack range measured in thousands of kilometers. While able to offer the ranges needed for global coverage from as few to as many as two dozen satellites, the SBL is unable to penetrate the earth's atmosphere. Ground- or air-based laser systems have to contend with the optical effects of passing through a medium of varying density to reach space targets. The passage of the laser itself caused atmospheric distortion as some energy superheats the air, hence the need to produce light on a wavelength that minimizes atmospheric absorption. Adaptive optics, computer controlled mirrors and optical windows able to detect and compensate for atmospheric distortion are prominent among other measures used to increase laser ranges.

Development for future weaponized lasers includes work on raising the power output of electrically powered solid state lasers into the megawatt range. An electrically powered laser would reduce the logistics of fielding a destructive effect laser weapon. Perhaps the ultimate form of weaponized laser technology is the free-electron laser (FEL); a device where the laser output is under a high degree of control over both power level and wavelength. The FEL uses magnets to manipulate an electron beam into producing a laser emission. Control over the electron beam and the powerful electromagnets allows for the output beam to be varied across a wide spectrum. The result is a very versatile and powerful laser, though weaponization of this technology lags far behind that of the chemical laser. Lasers powered by electricity require improvements in the efficiency of conversion of electricity to laser energy, and in power generation.

Consumer laser devices come with eye safety warnings, as they are capable of delivering enough light to cause temporary sight damage and permanent injury under some circumstances. This is also true for electro optical sensors found on military hardware and satellites. Recently there has been a rash of incidents where aircrews were dazzled by lasers. These include accidental exposures from laser light shows, and more malicious intentional incidents. While the effects have for the most part been temporary, 'dazzling' has put at risk aircraft, crew and passengers. During the Bill Clinton Presidency a U.S. laser sensor sparked controversy when it was used to illuminate a U.S. satellite as part of tests to assess satellite vulnerabilities. This test sparked fears over U.S. intentions to deploy an ASAT capability.[16] Recently the Chinese have been accused of using a satellite blinding capability against U.S. satellites, but like the previous example this may have just been another laser sensor. Then again, getting precise measurements of a foreign satellite via laser sensor could be more than just scientific research. The difference between dazzling and permanent blinding is only the matter of the laser's power output and there is no great leap in technology between the equipment needed to obtain both effects.

At much higher levels of power, a laser will rapidly heat the targeted area. The THEL laser program has demonstrated power levels capable of melting through thick artillery shells to set them off in midflight. ABL, which is now working toward its first airborne tests, aims to soften the fuselages of liquid fuelled ballistic missiles causing destructive structural failure. If a large enough quantity of energy is delivered quickly enough, the material, which absorbs the energy does not just melt, but instead explodes as the material rapidly transitions to gas. A potential spin-off from high-energy laser research is spacecraft propulsion, where a laser external to the spacecraft is used to rapidly heat and expand a propellant, expelling it at velocities beyond those possible through a combustion-based rocket. This propulsion concept can use practically any form of matter, gas, liquid or even solid, and capitalizes on the laser's potential to deliver large amounts of energy over great distances. Development of laser propulsion may work the other way and spin-on to form the basis of a weapon.

High-energy lasers are not necessarily the ultimate weapon for space and terrestrial applications. Once again the problem of power has thus far limited earthly applications to less-than-lethal eyesight and optical sensor dazzlers, which run afoul of international agreements and criticism even when steps are made to ensure effects are temporary. Just as military pilots can make use of eye protection (basically enhanced aviator sunglasses), satellite optical sensors can be equipped with shutters which protect optics when not in use, or provide a blink capacity able to close a shutter before permanent damage occurs. Physically destructive weapons can potentially be shielded by increasing the thickness of surface materials that may be exposed to laser attack. The problem with physical armor for ballistic missiles is the same as for satellites: it cuts into payload. Solid fuel rocket propulsion may give a ballistic missile superior resistance to laser attack due to differences in construction but even this is temporary.

Taking a nod from science fiction, laser energy may be dispersed by imposing some kind of deployable shield. Just as the atmosphere can absorb and block laser energy, satellites can be equipped with systems for deploying a solid or semisolid (aerosol foam or aerogel) barrier as needed. The shielding material would ablate away, carrying with it the energy of the attack in the process. The counter to this countermeasure would be to simply increase the energy delivered either with a more powerful laser or increasing the time a laser weapon dwells on the target. Moving beyond the application of spray-on shielding is the generation of a protective cloud of plasma (ionized gases), which could be lighter than spray-on ablative material. Cold plasma technology is also claimed to be capable of sheathing a vehicle from radar detection (though it is never explained as to how to eliminate the optical glow of plasma technology). Lasers and other DEW technology may bring along a revolution, but the race between defense and offence will not end.

Weapons based on particle beam technologies are also classified as directed energy weapons. The projected sub atomic particles are not considered solid matter. In a particle beam weapon, charged particles are accelerated to a sizeable fraction of the speed of light by magnetic and electrical fields. Though operating at below the speed of light, particle beams can easily deliver high energy by using heavier particles than the light photons of a laser. Charged particles can be steered electronically, as is done in old fashion cathode ray tube (CRT) televisions. Aiming the particle beam with fine manipulation of magnetic and/or electronic fields potentially speeds up the weapon's ability to move between targets, while increasing reliability through the elimination of mechanical components. The earth itself has a magnetic field, and is surrounded by trapped charged particles, meaning that use of a charged particle beam in proximity of the earth will result in the beam being deflected off course. In space weapons applications, the neutral particle beam was of some interest during the early part of the Strategic Defense Initiative. For the global ranges envisioned by SDI, the particle beam would be rendered electrically neutral once the particles were accelerated and aimed, so as to not face interference from the earth and other sources of EM.

Particle beam weapons were not pursued to the extent that KE and lasers were for near term deployment. The costs and risks associated with particle weapon development at the time were unfavorable compared to those for Brilliant Pebble kinetic energy interceptors, and the space-based laser concept. While it remains a further off technology, the particle beam still has potential as a direct energy weapon. In line with the current interest in control and focus over effects, the intensity of the particle beam can be controlled. At low power a particle beam can, like natural high-energy particles, cause temporary failure in electronics, or even rapidly expose a satellite to a life time's worth of radiation, raising the possibility that the satellite will completely and irreversibly fail due to natural causes. The same problems of damage assessment that apply to subtle HPRF attack apply to covert particle weapon use as well. At higher energy levels, a particle beam can deposit enough energy to cause physical damage to materials. Particle beams used in research for civilian applications can produce respectable sized craters in metal target blocks.[17]

Space-to-Surface Force Application

The debate over space weapons often hinges as much on where the weapon attacks and where the weapon is based as much as it considers its effects. Often ground-based weapons are left out of being categorized as a space weapon despite its effects occurring in space. This has allowed some degree of ambiguity for systems such as the Chinese ASAT tested in 2008, and U.S. land- and sea-based ballistic missile interceptors. Similarly weapons that transit through space fall into the same fuzzy grey area with respect to the space weapons debate. The inclusion of such systems in a comprehensive definition of space weapon then brings us to the reality that space has been directly exploited for warfare since the first Nazi V-2 entered space on its way to London. The clear cut example of a weapon that is permanently space based, in other words a weapon in orbit, brings up many technical arguments over the utility of space weapons. An examination of space-to-surface concepts highlights many of the obstacles facing near-term space weaponization, and the possible dysfunction of some concepts for so called space weapons.

In Stanley Kubrick and Arthur C. Clarke's *2001 A Space Odyssey,* an ancestor of humanity throws a bone, freshly used as a weapon against his own kind, into the air, which is tracked by the camera until the scene jumps to the near future (for the 1960s) and an orbiting nuclear weapons platform passes by. Metaphorically this represents that human nature has not changed over the millennia between prehistory and the future. With respect to space weapons, the orbiting nuclear weapons platform represents a bit of a technological dead end. Through the lens of the early cold war, the imagery of Kubrick's masterpiece makes sense; the space race is intertwined with the nuclear arms race, with the orbiting of Sputnik only a few years earlier certainly raising the specter of orbiting communist A-bombs. However, in hindsight the science of space has rendered the concept of orbiting weapons of mass destruction (WMD) as being a somewhat implausible threat.

Due to the mechanics of deorbiting a warhead an orbital surface strike platform would be uncompetitive verses ballistic missiles with respect to propulsion requirements. To be effective, there must be enough propulsion, or Delta-V, available to rapidly put a warhead on a trajectory that intersects the earth. A warhead that takes multiple orbits from launch to impact would take hours to conduct an attack exposing it to counterattack. In addition propulsion technology affects the number of platforms needed. Unless the surface strike platform is in GEO, it will only be in position to carry out an attack for a fraction of its orbit. GEO is a high orbit, which would compound the deorbiting problem. To increase the range at which an individual orbiting surface strike platform would be effective, more powerful propulsion must be attached to the warheads. This propulsion must, however, be put into orbit with the warheads, therefore increasing launch cost further. The economics of the Brilliant Pebbles constellation made of several thousand satellites was based on low cost satellite manufacture and expected low cost industrial small satellite launch. Nuclear weapons for many reasons are not low cost items, nor are they particularly light weight. Dispersing orbiting nuclear weapons into smaller and more numerous satellites worsens the problems associated with command and control. Nuclear weapons are not something that you want to lose control over. Placing nuclear weapons onto manned platforms only magnifies the already great financial burden of what is, arguably, already a questionable deterrent.

Compounding the ease at which a low orbiting satellite may be found (and attacked) it can be expected that an adversary would pay extreme attention to finding and monitoring orbiting nuclear weapons platforms. The existence of orbiting nuclear weapon strike platforms would be incentive for the deployment of ASAT systems. In a nuclear warfare mindset, ASATs would perhaps be less about securing space for ones own use, but instead neutralizing the others nuclear strike capabilities, with the consequence of perhaps a more relaxed attitude toward the use of messy kinetic energy and even nuclear ASATs. Warhead vulnerability also decreases a system's deterrent value. A vulnerable system under some nuclear strategies leads to a lower threshold for weapons use, the use-it-or-loose-it effect, which is considered destabilizing.

In comparison the terrestrial triad of nuclear weapons delivery systems, land- and sea-based missiles along with air delivered bombs and cruise missiles, offers greater survivability and flexibility for less cost. A long-range ICBM only needs enough thrust to throw its deadly payload to the desired range. The propulsion requirements of ballistic flight, even for global ranges, is significantly less than that needed to put the warheads into orbit along with the propulsion needed for rapid deorbit. The ballistic flight path of an ICBM only takes minutes to complete. On earth, missiles can be protected by building massive bunkers capable of surviving nearby nuclear detonations, or by placement on smaller mobile launch systems. Today missiles can be dispersed on railroad cars, on all-terrain trucks, and on near undetectable ballistic missile carrying submarines. Indeed the submarine launched ballistic missile is the only nuclear strike system maintained by the United Kingdom, which dispensed with bomber and ground-based ballistic missiles decades ago. Manned nuclear armed bombers provide flexibility over how an attack is conducted, including the possibility of recall. Bombers have kept up with interception technology with electronic warfare, stealth and standoff cruise missiles.

All of the nuclear delivery methods employed by the United States and the Soviet Union (now Russia) overlap to ensure survivability against a first strike. The nuclear triad is not a quartet because of the great cost involved with turning the orbit-to-surface concept into a credible deterrent. This of course did not stop development work from being conducted on orbiting WMD platforms; however, none have made it off the drawing board. Orbiting weapons of mass destruction are prohibited by the 1967 Outer Space Treaty, however, it would seem that by banning WMDs in orbit, not much utility was lost. Indeed it would appear that the banning of WMDs in orbit had more utility in the overall cold war struggle for public relations than it would have had it actually pursued been as a military capability.

The counterpoint to this is the Soviet deployment of WMD platforms based on the fractional bombardment (FOB) system. FOB lofted a nuclear warhead bearing missile bus into a very low orbit with the intention of warhead reentry prior to the completion of one orbit. The orbital trajectory is needed to carry the warhead halfway around the world, and if it were not for the intervention of missile bus' onboard propulsion and the air resistance present at such low orbit the FOB system would orbit the earth, in clear contravention of the 1967 OST. That being said FOB was classified by Soviet sources as an ICBM and therefore by their interpretation of the FOB system it was exempt from the 1967 OST. FOB had many of the characteristics of a long-range ICBM, not surprising as the deployed FOB was based on an ICBM. A small number of R-36-O FOB missiles were deployed briefly from 1969 to 1983.[18] The R-36-O was a variant of the R-36 heavy ICBM, which had a significantly larger payload or 'throw weight' than comparable U.S. ICBMs.

Compared to its ICBM brethren, the R-36-O only carried one warhead, compared to the many carried by the R-36 in the ICBM role. This highlights that orbiting requires more propulsion, and the need for payload to be devoted to deorbit propulsion. However, this one warhead could be launched in any direction to reach its target. The ability to deorbit at will also meant that the true target of FOB could be obscured, unlike an ICBM where the target could be reasonably estimated from its ballistic course. Moreover the FOB concept flew barely into space; minimizing the time the missile bus would be visible to ground-based sensors, and producing attack times comparable to an ICBM. Orbiting by definition gives FOB global attack range, but unlike a fully orbiting WMD platform, FOB had all the survivability benefits of a terrestrial system. In an odd way, Soviet paranoia over the U.S. space shuttle could be seen as a reflection of their own FOB deployment. With not much imagination the space shuttle program could have been viewed by the Soviets as a natural U.S. response to the R-36-O, a reusable fractional orbit bomber with a potentially larger throw weight.[19]

FOB demonstrated that while there are serious obstacles and costs associated with orbit to surface strike, these problems do not necessarily affect the whole concept of using space to apply force on the surface of the earth. While many of the problems with orbiting nuclear weapons apply to orbiting conventional surface strike weapons, many of the opportunities that FOB opened for nuclear attack, can also be applied to conventional strikes against targets on the earth. Given relatively minor advances in launch vehicle technology, the FOB concept could make use of any of the terrestrial basing methods already in use by nuclear weapons and take advantage of the benefits each has to offer avoiding the vulnerabilities of orbital basing. Suborbital bombardment beyond the already proven ballistic missile is still a strong contender for the global strike mission.

Today in the United States there is interest in developing a prompt global strike capability. The United States already has global reach via its aircraft carriers, and long-range air refueled bombers. However, these strike systems take time to get into position, and they require the consent for over flight by any country that sits between the United States and the target. Passing through space or at extremely high altitudes would allow an aerospace craft to reach easily the hypersonic speeds already used by ICBMs and launch vehicles. Long duration hypersonic cruise allows an aerospace craft to reach any point on earth from the continental United States. Flight though space, or at altitudes where sovereign airspace is debatable, avoids the need to obtain over flight permission from countries that stand in between the United States and the target. The process for obtaining permission to fly through airspace, even that of an allied nation can be as long as the actual mission itself. Moreover, allies can, and do, deny use of airspace. Candidate systems for the prompt global strike mission include land- and sea-based ballistic missiles converted to a conventional strike role, manned hypersonic bombers, sub orbital systems and even orbital systems.

ICBMs and SLBMs offer a proven solution, and are often just existing U.S. Air Force Minuteman or U.S. Navy Trident missiles with their nuclear payloads replaced by some kind of precision guided weapon. A criticism of this approach is that by using an existing nuclear warfare system, the use of a conventionally armed Minuteman or Trident missile could be misinterpreted as a U.S. nuclear attack by a third-party nuclear power otherwise not involved in the conflict, provoking a retaliatory nuclear strike against the United States. That being said, retasking existing Minuteman and Trident missiles for conventional strike, even with the cost of developing a precision reentry vehicle is currently the lowest cost option available.

The prompt global strike concept was developed in a U.S. defense establishment committed to precision attack. The common aero vehicle (CAV), or X-41 under the U.S. X-series of research aerospace craft, is part of DARPA's FALCON program. The CAV takes advantage of the atmosphere for maneuver, using aerodynamic control to make rapid and significant course changes, something impossible for an orbiting spacecraft to achieve with near term propulsion technology. Of course, the CAV is not traveling at orbital speeds, though this fact in itself only highlights that orbiting can hinder the application of space technology to surface attack. The U.S. space shuttle has already demonstrated a large cross range as it glides back to earth at hypersonic speeds. Some nuclear warhead reentry vehicles also have a bit of glide capability, but not to the degree of either the space shuttle or the proposed CAV. Once the CAV has neared its target, it is meant to release a variety of conventional guided munitions or surveillance equipment. The CAV is essentially a precision hypersonic munitions dispenser. To facilitate this role, it is hoped that the price of fielding a CAV would be comparable to that of a cruise missile.[20]

Aside from rocket powered launch vehicles and a weaponized X-37 space plane (which itself is launched by rocket power), hypersonic aircraft using the air-breathing scramjet are also suggested for boosting CAV-type munitions carriers to suborbital speeds. At the time of writing scramjets are still in the research stage, though at the very least they have been proven to function in real life under the X-43A program and the HyShot program conducted by the University of Queensland, Australia, in partnership with several international partners. Under the FALCON program is the Hypersonic Cruise Vehicle (HCV), a program to build an aircraft capable of hypersonic speeds approaching orbital velocities.[21] In contrast to the X-43A and

HyShot, the HCV will be takeoff and land from a runway, instead of being boosted by a rocket (the X-43A and its booster rocket are also air-launched from a B-52 bomber). Scramjets require boosting to high supersonic Mach numbers (approx Mach 4) to function, and part of the HCV program is the development of engine technologies that can take an aircraft from a standstill to speeds that overlap those at which a scramjet will operate. Integrating both low speed propulsion and scramjets represent a daunting challenge, and has been attempted before, notably in the defunct orbit capable X-30 program. Compared to the X-30, the goal of a global range suborbital aircraft or disposable cruise missile is much more modest.

An orbit to surface conventional strike is, however, not a completely shelved concept. There are still proponents for orbital basing. Expanding on suborbital CAV deployment, the same technology is being suggested in an orbit-based version, though in a more robust form in order to survive reentering the atmosphere at orbital speeds. The significant energy embodied by an orbit body can be converted to destructive power through the use of kinetic energy bombardment. A KE strike system can be as simple as a rod of material that can survive reentry reasonably intact (such as tungsten or depleted uranium) propelled toward the earth, with perhaps some kind of maneuvering capability to make fine adjustments to the course as it streaked through the atmosphere at nearly 25 times the speed of sound. This concept is often referred to under the 'Rods from God' nickname.

If a space transiting bomber or munitions dispenser can be developed, then the follow-on is the delivery of soldiers, the smartest of all weapons, via space. Spaceborne troop delivery is not a new concept in the United States, and has been entertained in aerospace circles for decades. Prominent among these were single stage launch vehicle concepts by engineer Philip Bono capable of hauling to anywhere on the globe hundreds of infantry plus equipment. Developed around the time of the Apollo moon landings, it only seemed natural that the next heavy lift rocket developed in the course of the space age could be used to insert a battalion to any trouble spot on earth. The space age has not developed in the manner expected in the 1960s; however, suborbital troop deployment has found new life leveraging off of the progress made toward affordable space tourism. In addition to the propulsion technologies that may allow spaceflight to fulfill the prompt global strike mission, troop insertion through space will require reliable life support. The technology needed to provide a reasonable level of safety and comfort for paying passengers will be more than sufficient for elite military forces, that routinely accept great risk.

Interest in prompt global access via space is not confined to the military. Rapid intercontinental cargo shipping via suborbital flight is today one of the usual future applications claimed by proponents of spaceflight technology development. The price and dangers of space flight remains barriers to the space delivery business plan. That being said, space tourism became a reality only a few years ago via the right combination of money and willingness to provide the service. Space borne delivery of cargo, munitions and special operations teams may very well follow in the wake of spaceflight holidays.

Military Space and Force Application II

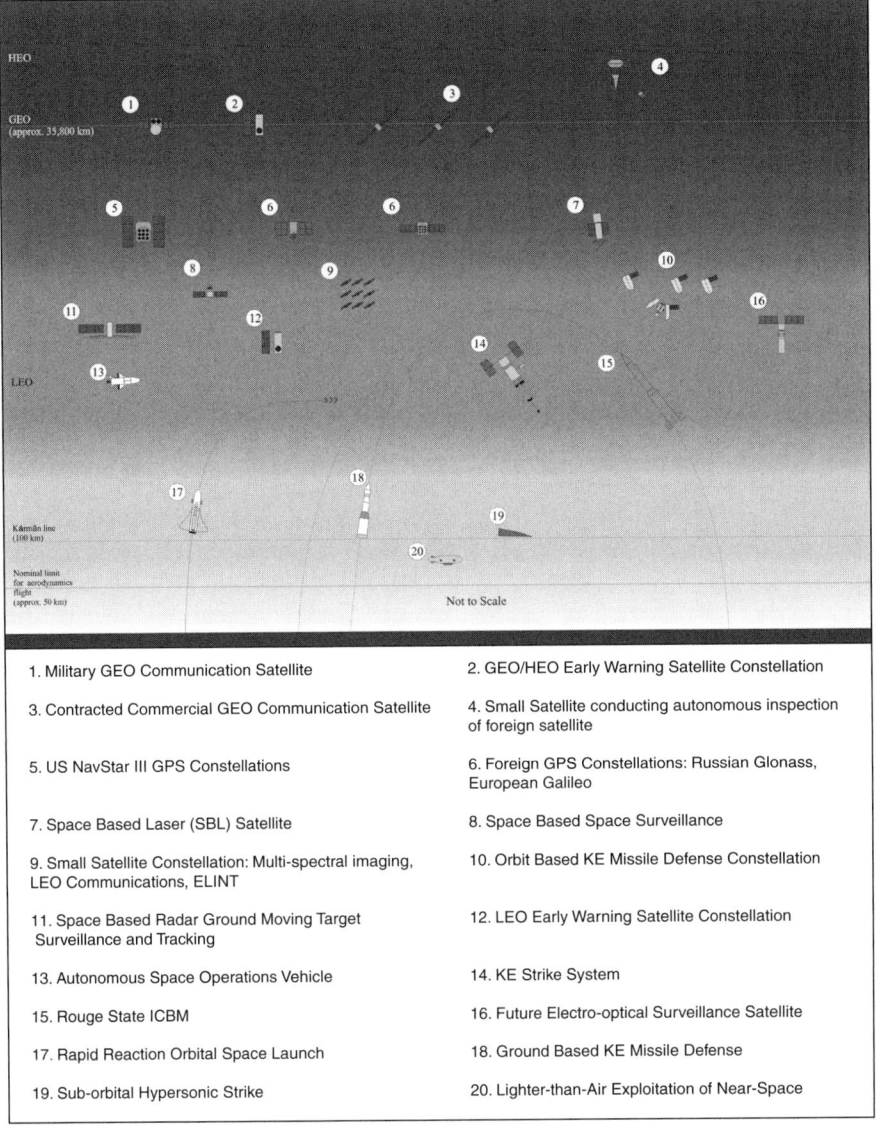

1. Military GEO Communication Satellite	2. GEO/HEO Early Warning Satellite Constellation
3. Contracted Commercial GEO Communication Satellite	4. Small Satellite conducting autonomous inspection of foreign satellite
5. US NavStar III GPS Constellations	6. Foreign GPS Constellations: Russian Glonass, European Galileo
7. Space Based Laser (SBL) Satellite	8. Space Based Space Surveillance
9. Small Satellite Constellation: Multi-spectral imaging, LEO Communications, ELINT	10. Orbit Based KE Missile Defense Constellation
11. Space Based Radar Ground Moving Target Surveillance and Tracking	12. LEO Early Warning Satellite Constellation
13. Autonomous Space Operations Vehicle	14. KE Strike System
15. Rouge State ICBM	16. Future Electro-optical Surveillance Satellite
17. Rapid Reaction Orbital Space Launch	18. Ground Based KE Missile Defense
19. Sub-orbital Hypersonic Strike	20. Lighter-than-Air Exploitation of Near-Space

Figure 4.2 Future Military Space

Conclusion

Cost-to-benefit is perhaps the best decider over whether specific space force application technologies are pursued into operation, and ultimately whether space militarization moves beyond the current fuzzy state of weaponization. Directly and indirectly technologies beneficial to active space force application seems to be making constant, though often inconsistent, progress. The sciences behind all the discussed space weapons concepts are sound; however, the engineering is not yet practical for many of them. Research and development costs must be weighed against a variety of factors: the threat situation, public finances, and the costs of competing space force application concepts. Promising, yet difficult technologies face competition from less capable but more within reach weapons. Not all systems are equal, and therefore some, if not most of what has been discussed may be superseded by events. However, every now and again, science fiction does become reality, and the technologies of the imagination become commonplace.

It cannot be ignored that there are political costs and benefits associates with particular technologies. How costs and benefits are weighted is a product of the politics of the time. At present there appears to be no rush to progress space force application beyond its relatively low key status. Indeed there are fears that increasing the space force application capabilities of the United States could jeopardize already proven space force enhancement capabilities by encouraging others to follow suit. On the other hand, the success of space force enhancement by itself may encourage others to find ways to neutralize it. More dangerously the technology of space force application is spreading. As the time of writing, North Korea, a pariah nation largely considered a danger to international peace and security, is laying claim to the right to have space technology along with its ambitions toward nuclear weapons. Peace it would seem is often more far off and unobtainable than some of the technologies discussed here. The technology of war in space as informed by political and strategic interests will remain a concern for some time.

Notes

1. Department of State, "Treaty on Principles Governing the Activities of States in the Exploration and Use of Outer Space, Including the Moon and Other Celestial Bodies," January 27, 1967, http://www.state.gov/t/ac/trt/5181.htm.

2. Ibid.

3. Craig Covault, "Chinese Test Anti-Satellite Weapon," *Aviation Week & Space Technology,* January 17, 2007, http://www.aviationweek.com/aw/generic/story_generic.jsp?channel= awst&id=news/CHI01177.xml.

4. Shirley Kan, *China's Anti-Satellite Weapon Test,* Congressional Research Service Report for Congress, April 23, 2007, http://www.dtic.mil/cgi-bin/GetTRDoc?AD=ADA468025& Location=U2&doc=GetTRDoc.pdf.

5. "Draft Treaty for the Prevention of Placement of Weapons in Outer Space," February 12, 2008, http://www.ln.mid.ru/brp_4.nsf/e78a48070f128a7b43256999005bcbb3/0d 6e0c64d34f8cfac32573ee002d082a?OpenDocument.

6. A monopropellant is a single compound that, on command, can break down chemically releasing energy, which in turn may be used to expel the by-products of the chemical reaction generating thrust in the process.

7. A variant of the nuclear-tipped Nike Zeus was an antisatellite weapon (ASAT) program in competition with Project 437.

8. Walter L. Perry and Marc Dean Millot, *Issues from the 1997 Army After Next Winter Wargame,* http://www.rand.org/pubs/monograph_reports/MR988/.

9. Congressional Budget Office, *Alternatives for Boost-Phase Missile Defense,* July 2004, http://www.cbo.gov/doc.cfm?index=5679.

10. David S. F. Portree, "NASA, Mir Hardware Heritage," March 1995, http://ston.jsc.nasa.gov/collections/TRS/_techrep/RP1357.pdf.

11. The kinetic energy weapon portion of the now suspended XM29 individual weapon program is at its heart a Heckler and Koch G36 based selective fire rifle firing the NATO standard 5.56 mm round.

12. National Institute of Standards and Technology, "CODATA Value: Speed of Light in Vacuum," http://physics.nist.gov/cgi-bin/cuu/Value?c.

13. Raytheon, "Silent Guardian" online sales brochure, http://www.raytheon.com/capabilities/products/stellent/groups/public/documents/content/cms04_017939.pdf.

14. Ibid.

15. Carlo Kopp, "The Electromagnetic Bomb—a Weapon of Electrical Mass Destruction," *Air & Space Power Journal,* 1996, http://www.airpower.maxwell.af.mil/airchronicles/cc/apjemp.html.

16. Arms Control Association, "U.S. Test-Fires 'MIRACL' at Satellite Reigniting ASAT Weapons Debate," *Arms Control Today,* October 1997, http://www.armscontrol.org/act/1997_10/miracloct.

17. Among the memories of a different academic career, one of the authors of this book had a professor who passed around the class Engineering EM Theory, a target block from his own research work.

18. Global Security, "R-36-O / SL-X-? FOBS," http://www.globalsecurity.org/wmd/world/russia/r-36o.htm.

19. Paul A. Czysz and Claudio Bruno, *Future Spacecraft Propulsion Systems: Enabling Technologies for Space Exploration* (Berlin: Springer, 2006), 46.

20. Terry H. Phillips, *A Common Aero Vehicle (CAV) Model, Description, and Employment Guide,* January 27, 2003, http://www.dtic.mil/dticasd/sbir/sbir041/srch/af031a.doc.

21. Defense Advanced Research Projects Agency: Tactical Technology Office, "Falcon," February 18, 2008, http://www.darpa.mil/tto/programs/Falcon.htm.

APPENDIX

Treaty on Principles Governing the Activities of States in the Exploration and Use of Outer Space, Including the Moon and Other Celestial Bodies

Signed at Washington, London, Moscow, January 27, 1967
Ratification advised by U.S. Senate April 25, 1967
Ratified by U.S. President May 24, 1967
U.S. ratification deposited at Washington, London, and Moscow October 10, 1967
Proclaimed by U.S. President October 10, 1967
Entered into force October 10, 1967

The States Parties to this Treaty,
Inspired by the great prospects opening up before mankind as a result of mans entry into outer space,
Recognizing the common interest of all mankind in the progress of the exploration and use of outer space for peaceful purposes,
Believing that the exploration and use of outer space should be carried on for the benefit of all peoples irrespective of the degree of their economic or scientific development,
Desiring to contribute to broad international co-operation in the scientific as well as the legal aspects of the exploration and use of outer space for peaceful purposes,
Believing that such co-operation will contribute to the development of mutual understanding and to the strengthening of friendly relations between States and peoples,
Recalling resolution 1962 (XVIII), entitled "Declaration of Legal Principles Governing the Activities of States in the Exploration and Use of Outer Space," which was adopted unanimously by the United Nations General Assembly on 13 December 1963,

Recalling resolution 1884 (XVIII), calling upon States to refrain from placing in orbit around the Earth any objects carrying nuclear weapons or any other kinds of weapons of mass destruction or from installing such weapons on celestial bodies, which was adopted unanimously by the United Nations General Assembly on 17 October 1963,

Taking account of United Nations General Assembly resolution 110 (II) of 3 November 1947, which condemned propaganda designed or likely to provoke or encourage any threat to the peace, breach of the peace or act of aggression, and considering that the aforementioned resolution is applicable to outer space,

Convinced that a Treaty on Principles Governing the Activities of States in the Exploration and Use of Outer Space, including the Moon and Other Celestial Bodies, will further the Purposes and Principles of the Charter of the United Nations,

Have agreed on the following:

Article I

The exploration and use of outer space, including the moon and other celestial bodies, shall be carried out for the benefit and in the interests of all countries, irrespective of their degree of economic or scientific development, and shall be the province of all mankind.

Outer space, including the moon and other celestial bodies, shall be free for exploration and use by all States without discrimination of any kind, on a basis of equality and in accordance with international law, and there shall be free access to all areas of celestial bodies.

There shall be freedom of scientific investigation in outer space, including the moon and other celestial bodies, and States shall facilitate and encourage international co-operation in such investigation.

Article II

Outer space, including the moon and other celestial bodies, is not subject to national appropriation by claim of sovereignty, by means of use or occupation, or by any other means.

Article III

States Parties to the Treaty shall carry on activities in the exploration and use of outer space, including the moon and other celestial bodies, in accordance with international law, including the Charter of the United Nations, in the interest of maintaining international peace and security and promoting international co-operation and understanding.

Article IV

States Parties to the Treaty undertake not to place in orbit around the Earth any objects carrying nuclear weapons or any other kinds of weapons of mass destruc-

Appendix

tion, install such weapons on celestial bodies, or station such weapons in outer space in any other manner.

The Moon and other celestial bodies shall be used by all States Parties to the Treaty exclusively for peaceful purposes. The establishment of military bases, installations and fortifications, the testing of any type of weapons and the conduct of military maneuvers on celestial bodies shall be forbidden. The use of military personnel for scientific research or for any other peaceful purposes shall not be prohibited. The use of any equipment or facility necessary for peaceful exploration of the Moon and other celestial bodies shall also not be prohibited.

Article V

States Parties to the Treaty shall regard astronauts as envoys of mankind in outer space and shall render to them all possible assistance in the event of accident, distress, or emergency landing on the territory of another State Party or on the high seas. When astronauts make such a landing, they shall be safely and promptly returned to the State of registry of their space vehicle.

In carrying on activities in outer space and on celestial bodies, the astronauts of one State Party shall render all possible assistance to the astronauts of other States Parties.

States Parties to the Treaty shall immediately inform the other States Parties to the Treaty or the Secretary-General of the United Nations of any phenomena they discover in outer space, including the Moon and other celestial bodies, which could constitute a danger to the life or health of astronauts.

Article VI

States Parties to the Treaty shall bear international responsibility for national activities in outer space, including the Moon and other celestial bodies, whether such activities are carried on by governmental agencies or by non-governmental entities, and for assuring that national activities are carried out in conformity with the provisions set forth in the present Treaty. The activities of non-governmental entities in outer space, including the Moon and other celestial bodies, shall require authorization and continuing supervision by the appropriate State Party to the Treaty. When activities are carried on in outer space, including the Moon and other celestial bodies, by an international organization, responsibility for compliance with this Treaty shall be borne both by the international organization and by the States Parties to the Treaty participating in such organization.

Article VII

Each State Party to the Treaty that launches or procures the launching of an object into outer space, including the Moon and other celestial bodies, and each State Party from whose territory or facility an object is launched, is internationally liable for damage to another State Party to the Treaty or to its natural or juridical persons by such object or its component parts on the Earth, in air space or in outer space, including the Moon and other celestial bodies.

Article VIII

A State Party to the Treaty on whose registry an object launched into outer space is carried shall retain jurisdiction and control over such object, and over any personnel thereof, while in outer space or on a celestial body. Ownership of objects launched into outer space, including objects landed or constructed on a celestial body, and of their component parts, is not affected by their presence in outer space or on a celestial body or by their return to the Earth. Such objects or components parts found beyond the limits of the State Party to the Treaty on whose registry they are carried shall be returned to that State Party, which shall, upon request, furnish identifying data prior to their return.

Article IX

In the exploration and use of outer space, including the Moon and other celestial bodies, States Parties to the Treaty shall be guided by the principle of co-operation and mutual assistance and shall conduct all their activities in outer space, including the Moon and other celestial bodies, with due regard to the corresponding interests of all other States Parties to the Treaty. States Parties to the Treaty shall pursue studies of outer space, including the Moon and other celestial bodies, and conduct exploration of them so as to avoid their harmful contamination and also adverse changes in the environment of the Earth resulting from the introduction of extraterrestrial matter and, where necessary, shall adopt appropriate measures for this purpose. If a State Party to the Treaty has reason to believe that an activity or experiment planned by it or its nationals in outer space, including the Moon and other celestial bodies, would cause potentially harmful interference with activities of other States Parties in the peaceful exploration and use of outer space, including the Moon and other celestial bodies, it shall undertake appropriate international consultations before proceeding with any such activity or experiment. A State Party to the Treaty which has reason to believe that an activity or experiment planned by another State Party in outer space, including the Moon and other celestial bodies, would cause potentially harmful interference with activities in the peaceful exploration and use of outer space, including the Moon and other celestial bodies, may request consultation concerning the activity or experiment.

Article X

In order to promote international co-operation in the exploration and use of outer space, including the Moon and other celestial bodies, in conformity with the purposes of this Treaty, the States Parties to the Treaty shall consider on a basis of equality any requests by other States Parties to the Treaty to be afforded an opportunity to observe the flight of space objects launched by those States.

The nature of such an opportunity for observation and the conditions under which it could be afforded shall be determined by agreement between the States concerned.

Article XI

In order to promote international co-operation in the peaceful exploration and use of outer space, States Parties to the Treaty conducting activities in outer space, including the Moon and other celestial bodies, agree to inform the Secretary-General of the United Nations as well as the public and the international scientific community, to the greatest extent feasible and practicable, of the nature, conduct, locations and results of such activities. On receiving the said information, the Secretary-General of the United Nations should be prepared to disseminate it immediately and effectively.

Article XII

All stations, installations, equipment and space vehicles on the Moon and other celestial bodies shall be open to representatives of other States Parties to the Treaty on a basis of reciprocity. Such representatives shall give reasonable advance notice of a projected visit, in order that appropriate consultations may be held and that maximum precautions may be taken to assure safety and to avoid interference with normal operations in the facility to be visited.

Article XIII

The provisions of this Treaty shall apply to the activities of States Parties to the Treaty in the exploration and use of outer space, including the Moon and other celestial bodies, whether such activities are carried on by a single State Party to the Treaty or jointly with other States, including activities are carried on by a single State Party to the Treaty or jointly with other States, including cases where they are carried on within the framework of international intergovernmental organizations.

Any practical questions arising in connection with activities carried on by international inter-governmental organizations in the exploration and use of outer space, including the Moon and other celestial bodies, shall be resolved by the States Parties to the Treaty either with the appropriate international organization or with one or more States members of that international organization, which are Parties to this Treaty.

Article XIV

1. This Treaty shall be open to all States for signature. Any State which does not sign this Treaty before its entry into force in accordance with paragraph 3 of this article may accede to it at any time.
2. This Treaty shall be subject to ratification by signatory States. Instruments of ratification and instruments of accession shall be deposited with the Governments of the United States of America, the United Kingdom of Great Britain and Northern Ireland and the Union of Soviet Socialist Republics, which are hereby designated the Depositary Governments.
3. This Treaty shall enter into force upon the deposit of instruments of ratification by five Governments including the Governments designated as Depositary Governments under this Treaty.
4. For States whose instruments of ratification or accession are deposited subsequent to the entry into force of this Treaty, it shall enter into force on the date of the deposit of their instruments of ratification or accession.
5. The Depositary Governments shall promptly inform all signatory and acceding States of the date of each signature, the date of deposit of each instrument of ratification of and accession to this Treaty, the date of its entry into force and other notices.
6. This Treaty shall be registered by the Depositary Governments pursuant to Article 102 of the Charter of the United Nations.

Article XV

Any State Party to the Treaty may propose amendments to this Treaty. Amendments shall enter into force for each State Party to the Treaty accepting the amendments upon their acceptance by a majority of the States Parties to the Treaty and thereafter for each remaining State Party to the Treaty on the date of acceptance by it.

Article XVI

Any State Party to the Treaty may give notice of its withdrawal from the Treaty one year after its entry into force by written notification to the Depositary Governments. Such withdrawal shall take effect one year from the date of receipt of this notification.

Article XVII

This Treaty, of which the English, Russian, French, Spanish and Chinese texts are equally authentic, shall be deposited in the archives of the Depositary Governments. Duly certified copies of this Treaty shall be transmitted by the Depositary Governments to the Governments of the signatory and acceding States.

IN WITNESS WHEREOF the undersigned, duly authorized, have signed this Treaty.

DONE in triplicate, at the cities of Washington, London and Moscow, this twenty-seventh day of January one thousand nine hundred sixty-seven.

GLOSSARY

Military-Related Terms, Activities, Events and Technologies

Advanced Extremely High Frequency (AEHF)/MILSTAR III	U.S. military communication satellite program to provide increased bandwidth, more security, and survivability for forces. Expected to be operational in 2010.
aerodynamic flight	Flight through lift generated by movement of an airfoil (wing or lifting body). Movement of a lifting generating shape causes air across one side to speed up versus that on the other, generating lower pressure on that side of the airfoil and consequently lift.
Air Launched Miniature Homing Vehicle (ALMHV)	U.S. ASAT program from the 1970s to the 1980s. A small kinetic energy interceptor (the miniature homing vehicle, or MHV) and its multistage suborbital launch vehicle launched by an F-15 into the path of an orbiting target where the MHV would find and lock onto its target via infrared optical sensors and conduct terminal maneuvers to ensure that the MHV collided with its target.
Ansari X-Prize	Private competition to encourage the development of space technology by the private sector (prize rules forbid government sponsorship). Formerly X-Prize (1996–2004), renamed Ansari X-Prize after significant donations by members of the Ansari family. Prize was won by Scaled Composites' Tier One Project (Spaceship One) on September 29, 2004, after completing two flights within time (two weeks), reusability (except for propellant, 90%), and payload (three adults) criteria to space (defined by the Kármán line, 100 km above sea level).

Anti-Ballistic Missile (ABM) Treaty	A 1972 treaty as amended by 1974 protocol limiting the Soviet Union and the United States to a single, 100 interceptor, ground-based missile defense around either the national capital or an intercontinental ballistic missile (ICBM) field. The Soviet Galosh system around Moscow remains operational. The American Safeguard system around the Grand Forks ICBM field was shut down in 1975. The treaty ended with the United States' notification of withdrawal in December 2001.
ASAT	Antisatellite weapon.
bent pipe	A term used in telecommunications to refer to how two ground stations without line of sight to each other use a satellite (typical GEO communication satellite) to cover long distances. A signal is transmitted from one ground station to the satellite, which retransmits the signal to the receiving ground station without further processing.
Bogotá Declaration (1976)	The formal claims of several equatorial states (Ecuador, Colombia, Brazil, Congo, Zaire, Uganda, Kenya, and Indonesia) to sovereignty over the geostationary orbit slot directly above their airspace. Their intention is not to eliminate satellites located at these points, but simply to charge rent.
boost phase intercept	The ideal missile defense intercept, which targets a missile during its initial launch phase when it is most vulnerable (a single large, slow target that is easily detectable by its heat signature) as it rises from the ground prior to exiting the atmosphere and detaching from its boosters.
Brilliant Eyes/SBIRS-Low/Space Tracking and Surveillance System (STSS)	U.S. early warning and missile tracking satellite constellation program, meant to support ballistic missile defense, originally identified during the Strategic Defense Initiative (SDI) years. Space-based infrared (SBIRS) in low earth orbit is designed to provide cold temperature tracking and target discrimination of warheads during their midcourse phase for the cueing of intercepts. It will be a constellation of around two dozen satellites. It has been plagued by a variety of development problems and cost overruns.
Brilliant Pebbles	Proposed U.S. space-based kinetic energy antiballistic missile system in the SDI era. The Brilliant Pebbles

Glossary

	plan would have maintained a constellation of about four thousand independent microsatellite-sized interceptors in very low earth orbit (orbit life spans of less than two years).
budget, Delta-V (velocity)	The expected amount of maneuvering or Delta-V (change in velocity) capacity a satellite expends in its lifetime. Often a limiting factor in a satellite's life span.
Committee on the Peaceful Use of Outer Space (COPUOS)	United Nations' committee formally established in 1959 U.N. Resolution 1472 as an international forum on the use of space. At present it has two standing subcommittees: Scientific and Technical Subcommittee, and Legal.
constellation	The use of multiple satellites to provide continuous coverage. As one satellite moves from line of sight, another takes its place. For some applications, such as Global Positioning System (GPS), it is necessary to have line of sight with multiple satellites, resulting in large constellations often distributed among multiple orbital planes.
Co-orbital ASAT	Soviet Union ASAT program that involved the orbit and rapid rendezvous (less than 1 orbit in later tests) of an interceptor satellite with its target. On rendezvous with the target, a conventional explosive would throw hot metal at the target satellite. This program ran from the mid-1960s to at least the late 1980s, reportedly including 20 or so orbital test intercepts, the majority of which were deemed successful. Present status of ASAT program is unclear, as is whether the system was actively deployed during the later part of the cold war.
Cosmos 954	A nuclear powered Soviet satellite, which failed to be boosted to its disposal orbit and instead reentered the earth's atmosphere. Debris from Cosmos 954, including nuclear materials impacted northern Canada.
Counter Communication System (CCS)/ Countercom	A mobile U.S. offensive counterspace system meant to disrupt satellite communication signals. Initial operating capability was in 2004.
coverage	In the context of this discussion, the area on the earth's surface that may be serviced by a satellite at a given moment. See also *footprint*.

Defense Intelligence Agency (DIA)	U.S. intelligence organization under the Department of Defense, which provides intelligence resources to all branches of the military.
Defense Support Program (DSP)	A small constellation of infrared early warning satellites maintained in GEO for the purposes of missile early warning. The satellites feed into the Missile Correlation Center, NORAD. Primarily designed for long-range ballistic missile threats, it proved successful in detecting the launch of SCUD missiles by the Hussein regime during the 1991 Gulf War.
defensive (bodyguard, sentinel) satellite	A satellite meant to protect another satellite from attack by potential ASAT weaponry. The distinction between *defensive* and *offensive* is for some controversial in that a system able to apply coercive force against an attacker can also apply coercive force against a target.
Delta-velocity (Delta-V)	A change in velocity. In the context of spaceflight, intentional Delta-V is produced by propulsion, either by rocket-type engines or through exotic and largely untried systems such as electrodynamic tethers, which use electromagnetic principles to obtain thrust or solar sails.
direct-ascent interception	Weapon flight profile where interception occurs at or near the apogee of a suborbital launch. Analogous to the flight of a research sounding rocket where the payload conducts it mission at the apogee of a suborbital launch. Used by U.S. SM-3 and ground-based interceptor (GBI) ballistic missile defense systems, as well as U.S. ALMHV and Project 437 antisatellite systems, as well as the People's Republic of China SC-19 ASAT.
directed energy weapon	A means of attack where destructive energy is not transmitted to a target by physical means (ramming, fragmentation, or concussion). Generally refers to laser-, radio-frequency-, particle-beam-, and electromagnetic-pulse-based weapons.
earth imaging (earth observation or remote sensing)	Using satellites to provide pictures of the earth's surface. Initially, this has referred to reconnaissance missions, but now includes all manner of civilian and commercial applications including mapping, forestry/agricultural surveys, environmental monitoring, and the sheer novelty of seeing one's house from orbit. Recently, there has been a growing trend of using com-

	mercially obtained earth imaging data for military use. This includes the use of commercial satellites to provide maps of theaters of war in both Afghanistan and Iraq, and Canadian plans to use commercial imagery (Radarsat) for polar and maritime sovereignty surveillance.
European Space Agency (ESA)	Intergovernmental agency coordinating the majority of space efforts among member nations and their space agencies. Canada has observer or partner status.
eXperimental Satellite System-11 (XSS-11)	A U.S. microsatellite program involving proximity operations. It has garnered some controversy over its potential to be easily adapted into an actual space weapon.
footprint	The area on the earth's surface that may be serviced by a satellite at a given moment. See also *coverage*.
Force Application and Launch from Continental United States (FALCON)	U.S. Defense Advanced Research Projects Agency hypersonic research program. Under FALCON are programs to develop a small launch vehicle to support research, and programs to construct and fly both rocket-launched and eventually runway-takeoff-and-landing hypersonic test vehicles.
fractional orbital bombardment (FOB)	Soviet-era strategic strike system, envisioned to bypass the United States' early warning radars directed northward by using an orbital trajectory (with an apogee well below that of the average ICBM) from a southern trajectory to approach the United States after almost completing a full orbit and present a shorter period of visibility to any horizon-limited sensors facing such an attack. FOB's orbit and deorbit characteristics gives it ideal first-strike or surprise attack capabilities. The status of such systems under the Outer Space Treaty (OST) ban on orbiting nuclear weapons remains to this day a matter of academic debate. Later arms-reduction treaties prohibited systems operating under the FOB concept.
Galaxy 4	On May 19, 1998, the Galaxy 4 satellite in GEO suffered a catastrophic failure in its altitude control system and backups, resulting in disruption of service to thousands of pagers, television, and gas station payment systems that suddenly lost their connection through space.

Galileo (GPS)	Originally a French military-security initiative and now a project of the European Space Agency (ESA), with the participation of China, to compete with the U.S. GPS system; the constellation will consist of 30 satellites in three orbital planes. A first test satellite was launched in 2005, with operational status projected for 2010.
Global Protections Against Limited Strikes (GPALS)	Post–cold war version of Regan-era SDI more suited for defense against a limited ICBM strike originating from the developing world than from a superpower-level nuclear exchange. Consequently, the number of ground- and space-based interceptors (Brilliant Pebbles) proposed is significantly lower. The plan may have included Russian participation.
GLONASS	Global Navigation Satellite System, the Soviet, now Russian, equivalent to U.S. NAVSTAR GPS constellation.
Ground-Based Electro-Optical Deep Space Surveillance (GEODSS)	U.S. optical telescope designed for space surveillance of ranges out to geostationary orbit. These large telescope sites are located in Socorro, New Mexico; Diego Garcia, British Indian Ocean Territory; and Maui, Hawaii. Morón, Spain, has a mobile telescope that contributes to the GEODSS system.
ground-based interceptor (GBI)	U.S. ballistic missile defense system. Each interceptor comprises a silo launched missile topped by a kinetic energy kill vehicle. Deployed in Alaska and potentially Poland.
ground station	Includes both ground control facilities and recipients of a satellite's services. In the case of GPS for example, U.S. satellite control facilities, handheld receivers, and receivers built into moving equipment (ships to guided munitions) would be counted as ground stations.
ground target moving indicator (GTMI)	Concept to use space-based radar to provide detection of moving targets on the ground.
Group of Eight (G8)	Informal grouping of eight of the world's leading industrialized nations that meet at an annual summit, originally to discuss economic matters. The G8 includes Canada, France, Germany, Italy, Japan, Russia, the United Kingdom, and the United States.
hardening	Shielding and other measures to protect a satellite. Every satellite is shielded in some manner against

Glossary

the space environment, in particular, natural radiation sources such as cosmic rays and the Van Allen belts (for satellites that spend time in that region). Basic shielding is generally insufficient to defend against the radiation effects of a nuclear detonation in space.

high ground, ultimate
In military (army) terms, the high ground bestows numerous advantages. It signifies the strategic importance of space, many space power advocates refer to space as the *ultimate high ground,* often with the suggestion that the control of space (earth orbital space) will lead to dominance in all other environments of human conflict.

hybrid launch vehicle
A multistage, partially reusable launch vehicle where the initial stage is reusable and later stages are expendable. The first stage does not reach orbital velocities and therefore has greatly reduced costs (both financial and technical) associated with recovery. This term is also used to describe a launch vehicle that uses hybrid rocket propulsion.

hybrid rocket
A rocket engine where one of the propellants is solid and the other is either a gas or a liquid. Typically, it is the fuel that is a solid with a liquid or gas oxidizer. At present, these types of rockets offer a trade-off in performance in favor of low cost and safety.

International Space Station (ISS)
The U.S.-led space station project for a range of space science purposes. Announced by President Reagan in the 1984 State of the Union address for completion within a decade, the first module was launched in 1998, and the station is not yet completed. It is currently more than 10 years behind schedule and double the original cost estimates of US$50 billion.

International Telecommunications Union
A United Nations' agency tasked with coordination of radio frequencies and the allocation of GEO orbital slots.

International Traffic in Arms Regulations (ITAR)
U.S. regulations controlling the import and export of defense and security sensitive technologies.

interoperability
The ability for different military agencies, and militaries from different countries, to work together cohesively. Requires not just hardware but also compatibility in force structures, doctrine, and training among other factors.

Iridium	Mobile telecommunications service based around a large constellation of 66 active satellites in low earth orbit (LEO). The Iridium service ground stations can be as small as oversized mobile phones.
Japan Aerospace Exploration Agency (JAXA)	The relatively new Japanese space agency, formed in 2003 from three previously separate entities.
joint direct attack munition (JDAM)	A type of U.S. precision guided munition using a combination of GPS and inertial navigation to achieve low-cost, all-weather, day/night precision strikes. Recently, a laser terminal seeker was added to allow for increased precision as long as weather and other conditions allow for laser guidance to be used.
Joint Space Project (JSP)	Bilateral program between Canada and the United States. For Canada it was a means of improving access to space-based information systems, such as communications and intelligence gathering.
joint tactical air-to-ground station (JTAGS)	U.S. system for direct in-theater warning of a ballistic missile attack and used to cue theater and tactical missile and air defense batteries such as the U.S. Patriot SAM/ABM batteries.
Kessler syndrome	Theorized situation where satellite destruction will lead to space debris, which leads to further satellite destruction and generation of more space debris in a cascading reaction. The result of the Kessler syndrome is that the density of space debris in stable orbits will render low earth orbit unusable for decades.
KeyHole (KH)	U.S. designation system for optical reconnaissance satellite systems. Imagery from earlier programs has been declassified for release.
kick motor/engine/stage	A rocket engine that on firing is meant to change the shape of a spacecraft's orbit, either from an initial parking orbit to a transfer orbit or from a transfer orbit into the final mission orbit.
kill vehicle	A weaponized spacecraft capable of maneuver to bring it into action against a target. Often used in context of a kinetic energy weapon.
kinetic energy weapon (KEW)	A weapon that employs the energy of an object in motion through collision to destroy another object. The space environment allows for extremely high speeds, meaning that the energies involved in a KEW attack

	are greater than that possible from the kill vehicle's equivalent mass in high explosives.
launch vehicle	A vehicle (rocket, missile) designed to move a payload into or through space. With respect to orbital space launch, it must be able to overcome gravity and accelerate a payload typically from a standstill to an initial orbital velocity needed for stable orbit.
launch vehicle, expendable	A launch vehicle that is used once with no effort made to recover it for further use.
launch vehicle, reusable	A launch vehicle that is recovered for reuse. Scaled Composites' fully reusable Space Ship One/Tier One is a suborbital vehicle only. The U.S. space shuttle, Soviet Buran/Energia system, and SpaceX's Falcon 1 are the only partially reusable orbital launchers to have attempted spaceflight. At present, there are no examples of fully reusable orbital launch vehicles.
Liability Convention (1972)	Formally, Convention on International Liability for Damage Caused by Space Objects, it elaborates on Article VII of the OST, which establishes responsibility for an object to the launching state who is liable for damage caused by space objects (satellites) on earth, land, sea, and in space. It also establishes procedures for settling compensation claims.
Limited Test Ban Treaty (1963)	Bilateral U.S.-Soviet treaty prohibiting all nuclear tests above ground, including in space and on the seabed.
Mach number	Ratio between the speed of an object and sound under the particular environmental conditions present for the object. Space does not present an adequate medium for sound transmission; hence there cannot be a speed of sound in space. Spacecraft velocities are often made in comparison to terrestrial Mach numbers (at sea level, 1,225 km/h), with Mach 25 being the necessary speed to reach a minimum orbit.
maritime economic zone	Offshore territory claimed by a nation for economic use.
midcourse intercept	Interception of a ballistic missile after its engines have finished firing and the missile payload (warhead) is coasting though to its apogee and onward to the reentry, or terminal phase, of ballistic missile flight.

Mid-Infrared Advanced Chemical Laser (MIRACL)	U.S. experimental laser facility located at the White Sands Missile Range, New Mexico. This is a megawatt-class chemical laser, which in the course of its use in experiments has demonstrated latent ASAT capabilities. In 1997 with much controversy over its weaponization potential, MIRACL was used to test the vulnerability of U.S. satellites to laser attack by briefly illuminating a soon to be out of service U.S. Air Force satellite.
MILSTAR	U.S. military communication satellite program. The AEHF (MILSTAR 3) program is the third generation of satellite to carry the MILSTAR name.
mission payload	Equipment carried by a satellite devoted to its mission.
Misty/AFP-731/US 51/ NSSDC ID: 1990–019B	Speculated U.S. reconnaissance satellite program incorporating stealth technology. AFP-731/US 51/ NSSDC ID: 1990–019B specifically refer to a U.S. satellite launched in 1990 by the space shuttle *Atlantis,* supposedly as part of the Misty program.
Moon Treaty (1979)	Formally, the Agreement Governing the Activities of States on the Moon or Other Celestial Bodies, it was meant to clarify a range of issues not fully dealt with in the 1967 Outer Space Treaty. It was never ratified and is not in effect.
National Aeronautics and Space Administration (NASA)	U.S. space agency responsible for civilian space exploration, among other missions or purposes.
National Missile Defense	Clinton-era U.S. antiballistic or ballistic missile defense program involving the use of ground-based midcourse interceptors. Its direct successor is the U.S. ground-based midcourse phase system operational as of 2006 at Fort Greely, Alaska, and Vandenberg Air Force Base, California.
National Oceanic and Atmospheric Administration (NOAA)	Part of the U.S. Department of Commerce and focused on weather science and weather prediction, operating numerous satellites.
National Reconnaissance Office (NRO)	U.S. agency that builds and operates reconnaissance satellites.
NAVSTAR GPS	U.S. Global Positioning System. NAVSTAR GPS provides not only free accurate positional data, but also an accurate timing signal on which most electronic financial transactions are synchronized. Accurate lo-

Glossary 133

	cation and accurate timing data also implies accurate motion data for moving objects equipped to receive GPS, allowing for its use in precision guided munitions (PGMs). It emits two signals; the less accurate public c/a code for general use and the encrypted p code for military use.
North American Aerospace Defense Command (NORAD)	Binational partnership between Canada and the United States, which has guarded North American airspace from intrusion since 1958. The agreement was renewed indefinitely in 2006.
North Atlantic Treaty Organization (NATO)	Based on the Treaty of Washington (1949), it is the Western collective defense organization originally designed to deter and defend Europe during the cold war.
open skies	Confidence-building measure for treaty verification purposes allowing for overflying of national territory by a state party to the treaty. Originally proposed by President Eisenhower in 1955. A formal treaty was negotiated between NATO and the then Warsaw Pact in 1992 and came into formal effect in 2002.
operational responsive space (ORS)	A movement to reduce the cost and time needed to identify and deploy a new space capability, usually in the context of space force enhancement. Technologies associated with ORS include microsatellites; low-cost, small-payload launch vehicles; and near-space aerial vehicles (as low-cost substitute for orbiting satellites).
orbit	To completely circle a body. The unpowered or free-flight state where baring other forces a spacecraft will circle the earth indefinitely due to a combination of gravity and the momentum imparted earlier by propulsion from a launch.
orbit, geostationary or geoearth (GEO)	A circular geosynchronous orbit aligned with the earth's equator giving a satellite an orbital period of one day, resulting in the satellite hovering above a point on the equator. An orbital altitude of about 35,800 kilometers is required to produce a day-long orbital period. Also called the Clarke orbit after author Arthur C. Clarke who popularized the concept.
orbit, geosynchronous (GSO)	Any circular orbit with an orbital period of one sidereal day. A satellite in a geosynchronous orbit will

pass over the same spot on its ground track at the same time every day. Geosynchronous orbits with an incline of near zero will from the surface of the earth seem to trace a figure-eight pattern confined to a relatively small region of sky overhead. A circular geosynchronous orbit with an incline of zero, where it is aligned with the equator, will have the satellite's ground track position matched with a point on the earth (see *orbit, geostationary*). Sometimes incorrectly used interchangeable with geostationary orbit. All unpowered geostationary orbits are geosynchronous, but not all geosynchronous orbits are geostationary.

orbit, high earth (HEO)	Any orbit beyond the nominal 35,800 kilometers geostationary orbital altitude.
orbit, low earth (LEO)	An earth orbit that is found between the altitudes of 150 kilometers and 2,000 kilometers above the earth's surface.
orbit, medium earth (MEO)	An earth orbit with an altitude between that of the upper limit of LEO (2,000 km) and below that needed for GEO orbit (approximately 35, 800 km). Also called an intermediate earth orbit.
orbit, Molniya	A highly elliptical orbit (perigee ~200 km, apogee ~40,000 km), with an inclination of 63.4 degrees or 116.6 degrees, and orbital period of about 12 hours, which gives the satellite extended coverage (on the order of 8 hours) over high latitudes.
orbit, polar	An orbit with an inclination of 90 degrees, resulting in an orbital plane that passes through earth's poles.
orbit, transfer	An elliptical orbit that intercepts both the original and the destination orbit. See also *kick motor/engine/stage*.
orbital apogee	The highest or farthest from earth point in an orbit.
Orbital Express	U.S. DARPA small satellite program involving a pair of satellites launched in March of 2007, Autonomous Space Transport Robotic Operations (ASTRO) and Next Generation Satellite/Commodity Spacecraft (NextSat/CSC). During the Orbital Express program ASTRO conducted proximity and docking operations with NextSat/CSC, including the transfer of fuel and components.
orbital inclination	The angle between the orbital plane and the equator. A geostationary orbit by definition must have an orbital inclination of 0 degrees.

Glossary

orbital perigee	The lowest or nearest to earth point in an orbit.
orbital period	The amount of time it takes for a satellite to complete one orbit.
orbital perturbations	Factors that cause an orbiting body not to follow the calculated orbit. It is a function of drag from the fringes of the earth's atmosphere at LEO altitudes, differences in gravity from the fact the earth is not the perfect sphere (used in calculations), third-party gravitational influences, and collisions.
orbital plane	An orbit can generally be described on a flat plane that bisects the earth.
Outer Space Treaty, 1967 (OST)	Formally, the Treaty on Principles Governing the Activities of States in the Exploration and Use of Outer Space, including the Moon and Other Celestial Bodies, it is the foundation of the international legal regime governing space. It establishes outer space, the moon, and other celestial bodies as international in character and for the use of all mankind. Specifically, space is treated as international waters, open to freedom of passage. It only prohibits the deployment of weapons of mass destruction (nuclear weapons primarily) in space, implicitly defined as an object completing a single orbit, and testing weapons, conducting maneuvers, or establishing military bases on the moon or other celestial bodies.
Palapa B1	Indonesian-owned Palapa B1 in 1996 directly jammed APSTAR-1A, a satellite owned by a Hong Kong–based company, during a dispute over use of an orbital slot. This infamous incident is widely cited as an early example of open conflict (attack) in space. Of particular note is that no major space powers were directly involved (Palapa B1 was U.S. built and launched, APSTAR-1A was U.S. built and Chinese launched, and Hong Kong was still British territory). Earlier in 1992, Palapa B1 challenged a satellite owned by U.S. company Rimsat for use of that very same GEO slot.
parasite spacecraft (space mine)	A satellite sent to orbit in formation with an opposing power's satellite, ready to conduct some action on command. Often in the context of ASAT spacecraft but can also be used in the form of less active means such as surveillance.

Powers, Francis Garry (1929–1977)	CIA U-2 pilot who was shot down in 1960 by Soviet SA-2 missiles (fired in salvo due to the United States reusing the same flight plan). Had to endure capture, a show trial, and 2 years of a 10-year sentence of hard labor, prior to release back to the United States. Later went on to work as a test pilot for Lockheed Martin's Skunk Works (manufacturer of the U-2). Posthumously awarded Prisoner of War Medal, Distinguished Flying Cross, and National Defense Service Medal.
precision guided munitions (PGMs)	Munitions with onboard terminal guidance capable of hitting exceedingly close to the aim point. Also referred to as *Smart* or *Brilliant* weapons.
Prevent an Arms Race in Outer Space (PAROS) committee	Ad hoc UN Conference on Disarmament committee, whose mandate is regularly proposed, though effectively blocked by the United States.
Program 437	U.S. direct-ascent ASAT program where a Thor intermediate-range ballistic missile was used to deliver a thermonuclear weapon to the vicinity of a target satellite as it passed overhead.
prompt global strike	The capability to strike any point on earth with non-nuclear weapons in time measured in hours and minutes rather than days. Candidate technologies for this as of yet unfulfilled requirement include hypersonic transatmospheric vehicles and maneuvering PGMs delivered by suborbital reentry vehicles.
Rapid Attack Identification Detection Reporting System (RAIDRS)	U.S. defensive counterspace program meant to identify type and source of interference with U.S. and friendly satellite operations.
Registration Convention (1976)	Formally, *Convention on Registration of Objects Launched into Space,* it requests that states maintain a national registry on objects launched into space and forward the registry to the United Nations.
Rescue Agreement (1968)	Formally, the *Agreement on the Rescue of Astronauts, the Return of Astronauts, and the Return of Objects Launched into Outer Space,* it requires state parties to facilitate the rescue and return of astronauts and objects.
responsive space launch	The capability to launch small payloads into orbit on demand. Related technologies include low-cost space access, space tourism, operationally responsive space, hybrid launch vehicles, air-launched launch vehicles, single-stage-to-orbit launch vehicles, and streamlining overall space access.

revolution in military affairs (RMA)	Often used to describe paradigm shifts in military power whether brought on by operational or technological changes. In the present context, refers to the information age doctrines and hardware, which it is argued has brought about a new way to fight wars.
RIM-161 standard missile 3	U.S. Navy medium-range antiballistic missile. SM-3 fits into the U.S. Navy Mk.41 vertical launch tubes, a multistage suborbital missile topped with a kinetic-kill vehicle. The direct-ascent SM-3 weapon is meant to carry out interception during the midcourse portion of the target missile's flight. Successfully adapted for an ASAT mission in 2008 to intercept the dead U.S. satellite USA 193.
rocket	A self-contained propulsion system that works by expelling propellant in a direction opposite to the direction of flight. At its most basic, a rocket is an implementation of Newton's third law of motion: for every action, there is an equal and opposite reaction. A rocket ejects via some kind of energy source a reaction mass in one direction, resulting in thrust in the opposite direction. In a chemical rocket, the propellants (oxidizer and fuel) both generate the energy and provide the reaction material. Thermal rockets have an external heat source (nuclear, solar, beamed energy, etc.) that provides energy to expel reaction mass. Electric rockets use electrical and/or magnetic principles to accelerate reaction mass, often in the form of charged particles.
satellite bus	Conceptually, in satellite design, the *chassis* or platform onto which mission equipment is mounted.
satellite telephony industry, rise and fall of	In the mid 1990s, there was great interest in establishing large constellations of low-orbiting communications satellites to service relatively (for the time) small handheld portable phones and other telecommunications. Competing technologies such as fiber optics and cellular (phone) networks proved more successful financially, leaving the satellite phone industry more or less defunct. Supporting the massive number of satellites envisioned were attempts at creating a commercially viable orbital space launch industry.
SC-19	U.S. designation for the Chinese ground-to-space direct-ascent ASAT. The SC-19 is believed to comprise

	a multistage solid fueled launch vehicle topped by a hit-to-kill kinetic energy interceptor. On January 11, 2007, the SC-19 system successfully intercepted a surplus but still functioning Chinese weather satellite, the FY-1C.
scramjet	Supersonic combustion ramjet.
segments	In space operations an entire space system is divided into conceptual *Segments:* space and ground. The space segment consists of the orbiting spacecraft. The ground segment is often divided between satellite control and users of the satellite's service.
Shenzhou	Chinese-manned spacecraft, similar to Soviet/Russian Soyuz but larger and with an orbital module capable of independent unmanned on-orbit maneuver/mission once the reentry capsule has separated for return to earth.
shutter control	In the context of this discussion, regulatory and legal means imposed on companies offering remote-imaging services to control the flow of such data to potential hostile powers.
signals intelligence (SIGINT)	In the context of this discussion ease dropping on electronic emission on the earth's surface via satellites, and on communications between satellites and ground stations.
small satellites	As computers become increasingly powerful and smaller, more capability can be fitted into smaller payloads. One result has been the development of smaller satellites, which may be further classified as follows:

- Minisatellite, 100–500 kg
- Microsatellite, 10–100 kg
- Nanosatellite, less than 10 kg
- Picosatellite, less than 1 kg

	There is no official or standardized terminology for classifying satellites by size.
space-based kinetic-kill vehicle (SBKKV)	Early SDI orbiting missile defense system. Hundreds of orbiting satellites would contain and provide initial guidance to several kinetic energy missiles. Concerns about survivability and cost led to this concept being shelved in favor of the Brilliant Pebbles concept.
space-based laser (SBL)	An orbiting antiballistic missile, high-energy laser weapon concept that is regularly proposed but as of

Glossary

	yet unfunded. A constellation of roughly two dozen orbiting laser weapon satellites could give constant coverage to a wide band of the earth.
Space-Based Space Surveillance (SBSS)	Proposed U.S. constellation of satellites for the purpose of monitoring objects in earth orbit. This program has reportedly had its schedule moved forward with the launch of the Pathfinder satellite moved up to 2008.
space-based visible (SBV) sensor	A sensor mounted on the Midcourse Space Experiment (MSX) satellite launched in 1996. At present, the SBV sensor is still functioning and is used to make observations on objects out to geostationary orbit distances for the U.S. Space Surveillance Network.
Space Command, United States (USSPACECOM)	From 1985 until 2002 when it was incorporated into United States, Strategic Command, U.S. Space Command was the unified (joint) command over U.S. military space operations.
space-faring	Space analogy to the maritime term *sea-faring* to describe a nation that has a certain competence in the various areas of space power. In this context, refers to a state that at a minimum has independent access (a national launch capability) to space.
space force application	Generally the application of coercive (military) force in space against orbiting targets, and from orbit against terrestrial targets. Midcourse missile defense systems by some definitions also qualify, as interception of an ICBM would occur in space. There are some schools of thought that also include weapons passing through space between terrestrial launch and target points (i.e., long-range ballistic missiles).
space force enhancement	The use of space-based assets to greatly enhance terrestrial warfare abilities in comparison to an unenhanced force. The use of space assets is said to have a *force multiplication* effect, allowing comparatively small forces to perform tasks that previously required large investments.
space force versus air/ aerospace force	Due to the importance of space to as a strategic realm, there are regularly suggestions that a space force be established as a separate branch of the military. The U.S. debate over the establishment of a separate space force has included the recent trend of the U.S. Air

Force describing itself as an aerospace force that presently emphasizes the air component of aerospace and will someday emphasize the space portion. Further compounding this debate are the parallels between space and naval warfare that have many science fiction authors envisioning space fleets and the term *space navy*.

space junk
: Orbiting artificial debris resulting from the various space programs around the world.

space lines of information (SLOI)
: Space analogy with maritime concept of sea lines of communication. Whereas the communications at sea refer to maritime commerce and transport of military force, the space medium provides for primarily the acquisition and movement of information, hence its usage in SLOI.

space militarization
: The use of space assets for military purposes. These range from cold war–era strategic surveillance to present-day force enhancement and, in the future, force application. It is currently distinguished from the weaponization of space, which by legal default consists of deploying a conventional weapon on-orbit.

space race
: A mostly peaceful competition between the West, led by the United States, and the Soviet Union to develop spaceflight. It occurred in the backdrop of the cold war. Arguably, it lasted from the Soviet launch of Sputnik in 1957 until the United States landed on the moon during the Apollo 11 mission of 1969. The term is being resurrected to describe both the race between Asia-Pacific powers (People's Republic of China, Japan, and India) to develop space systems and the potential competition between the United States and others (People's Republic of China, Russia, Europe, Japan, and India) to return a man to the moon in or around a nominal 2020 deadline.

space sanctuary
: Strategy where space is kept weapons free (ASAT free) through norms and agreements, so as to allow its use for strategic surveillance, early warning (in the context of nuclear war), and space force enhancement. Policy statements notwithstanding, due to limitations on technology, this is the de facto situation at present.

Glossary

space shuttle	Conceptually, this term is used to describe various real and proposed spacecraft that use wing born flight, usually for recovery. In general, it refers to the U.S. Space Transportation System, which is a partially reusable crewed launch vehicle.
space surveillance	In the context of this discussion, knowing what is in orbit that may present a hazard (i.e., space junk) or a threat (i.e., a satellite doing the bidding of a hostile force).
Space Surveillance Network	A collection of U.S. ground-based radar and telescope facilities plus one optical sensor based on an orbiting satellite that provide the United States with space surveillance.
Space Tracking and Surveillance System (STSS)	See *Brilliant Eyes*.
space weaponization	A rather contentious term, in general it refers to a situation where space force application is widely practiced by one or more nations. At present due to latent space weapons capabilities found in nuclear armed missiles (fused to go off at high altitude), space weapons experiments (such as MIRACL laser facility at White Sands, New Mexico), and the fact that ballistic missiles do transit through space, the distinction between today's state of space militarization and weaponization can be described as fuzzy.
staging, multi	To improve the payload and structure verses fuel mass ratio. Staging is used to shed dead weight (by dropping spent stages) as a launch vehicle ascends. Tsiolkovski's rocket equation provides the mathematical rational for multistage rockets.
staging, single	While staging offers relaxed mass fractions for payload, it is operationally expensive due to the complexity of having essentially multiple vehicles that all must operate correctly. For a rocket-powered single stage to orbit vehicle, it is not uncommon to find the propellant between 80% and 90% or more of the total launch mass. Air-breathing propulsion allows for lower propellant requirements as the oxidizer is largely supplied by the environment.

Starfire Optical Range (SOR)	A telescope facility located at Kirtland Air Force Base in New Mexico that is renowned for work on adaptive optics to compensate for atmospheric distortion. Is reputed to be able to make relatively detailed observations on orbiting spacecraft.
Starfish Prime	The United States test in 1962 of a 1.4 megaton warhead at an altitude of 248 miles over the Pacific Ocean, causing a blackout of communications over the area and permanently damaging three satellites in orbit. In 1963, such tests would be prohibited with the Limited Test Ban Treaty.
Strategic Air Command (SAC)	U.S. Air Force command that operated ICBM and bomber forces during the cold war. It is now part of U.S. Strategic Command.
Strategic Command (USSTRATCOM)	Current U.S. unified command, which among other missions is responsible for U.S. military space operations and U.S. nuclear deterrent forces (ICBM, bombers, and submarine-launched ballistic missiles). Just recently, two subordinate commands for each mission were established—Space and Global Strike.
Strategic Defense Initiative (SDI) or Star Wars	Ambitious U.S. ballistic missile defense research and development program, which included substantial space elements, including orbiting weapons. Proposed in 1983 by Ronald Regan and scaled back considerably by his successors, SDI led to many research programs that have a direct bearing on today's debate on limited missile defense and space weapons.
suborbital	A spacecraft that has at least achieved the nominal definition of space (the Kármán line at 100 km altitude) but does not have the momentum needed to complete one orbit and is pulled back down by gravity.
suborbital space (sometimes referred to as near-space)	Region between the limits of aerodynamic flight (approximately 50 km altitude) and the minimum altitude where an unpowered orbit will not rapidly decay (approximately 150 km). This region is often referred to as a no-man's land between aeronautics and spaceflight, as it very difficult to do more than simply traverse this area on rocket thrust. Very-high-altitude ballooning (in the lower parts of this region) and some transatmospheric vehicle technology (for the higher end) may allow exploitation of this domain.

Glossary

Teal Ruby	U.S. research program into satellite-based infrared sensors to be used for detection of aircraft from space. Canada and Australia participated in the project.
U.S. Army 1997 Winter War Game	In the U.S. Army's 1997 Winter War Game Army After Next, the BLUE Team, an RMA force of 2020, was brought to the negotiating table by the near-peer RED Team who detonated several nuclear weapons in space crippling the BLUE Team's ability to fight. The findings also estimated that the attack would have plunged the world economy into global depression for over a decade.
unmanned aerial vehicle (UAV)	In contemporary terms refers to semiautonomous aircraft. For beyond line-of-sight operations from ground controllers, UAVs depend on satellite communications to relay back data and to receive commands. GPS navigation is also important for UAV operations.
unmanned combat aerial vehicle (UCAV)	Armed UAV aircraft. While ad hoc antiarmor armed Predator UAVs may be considered UCAVs, this term generally is used to describe a dedicated autonomous warplane able to reach and engage targets with little human supervision.
USA 193	U.S. reconnaissance satellite that failed shortly after launch in December 2006. Due in large part to fears that its total load of hydrazine fuel would survive its expected reentry in 2008, an ad hoc plan to destroy the satellite in the weeks prior to its reentry was put together. On February 21, 2008, a single U.S. Navy SM-3 missile launched by USS *Lake Erie* was able to destroy USA 193.
Van Allen radiation belts	Two natural donut-like belts of highly charged particles formed by the earth's magnetic field redirecting radiation from natural sources. These belts are a design concern for space missions as prolonged exposure will degrade electronics and some materials. The size and intensity of the Van Allen belts is enhanced by natural solar activity and by man-made high-altitude nuclear detonations.
Vision for Space Exploration (VSE)	A 2004 U.S. policy concerning completion of the ISS, retirement of the space shuttles, shuttle replacement (spacecraft recently named *Orion*), return of crewed

expeditions to the moon, and eventual crewed exploration of the planet Mars.

X-41 common aero vehicle (CAV)
U.S. reentry vehicle project meant to allow for a wide degree of cross-range maneuvering. Reportedly it is meant for a suborbital launch and is under consideration for a U.S. prompt global strike capability.

X-ray laser (nuclear bomb pumped)
A weapons concept worked on under the U.S. Strategic Defense Initiative. Reportedly, it was to project multiple high-energy laser beams in multiple directions, all powered by the detonation of a nuclear device. Basing concepts included ground- and sea-based direct-ascent boosters, and on-orbit mines (though the use of a nuclear detonation as the power supply led to arguments against this basing concept due to possible conflict with the 1967 Outer Space Treaty).

X-37
U.S. autonomous space plane project that is ongoing. It is to be vertically launched by expendable multistate rockets and landed horizontally for reuse. First flight has been delayed from late 2008 to early 2009.

X-20 Dynasoar
U.S. manned space plane project that ran from 1957 to 1960. It was to be vertically launched by expendable multistate rockets and landed horizontally for reuse.

Bibliography

Primary

Ames Research Center, National Aeronautics and Space Administration. "Remote Agent Experiment." http://ic.arc.nasa.gov/projects/remote-agent/faq.html.

Barr, Larine. 88th Air Base Wing Public Affairs, United States Air Force. "Pulsed Detonation Engine Flies into History." *Air Force Print News Today,* May 16, 2008. http://www.afmc.af.mil/news/story_print.asp?id=123098900.

Commission to Assess United States National Security Space Management and Organization. *Report of the Commission to Assess United States National Security Space Management and Organization,* January 11, 2001. http://www.space.gov/docs/fullreport.pdf.

Congressional Budget Office. *Alternatives for Long-Range Ground-Attack Systems.* March 2006. http://www.cbo.gov/ftpdocs/71xx/doc7112/03-31-StrikeForce.pdf.

Defense Advanced Research Projects Agency: Tactical Technology Office. "Falcon," February 18, 2008. http://www.darpa.mil/tto/programs/Falcon.htm.

Department of Defense. *Report of Defense Science Board Task Force on High Energy Laser Weapon Systems Application,* June 2001. http://www.acq.osd.mil/dsb/reports/rephel.pdf.

Department of Defense. *Report of Defense Science Board Task Force on the Future of the Global Positioning System,* October 2005. http://www.acq.osd.mil/dsb/reports/2005-10-GPS_Report_Final.pdf.

Department of Defense. *Transformation Planning Guidance,* April 2003. http://www.defenselink.mil/brac/docs/transformationplanningapr03.pdf.

Department of State. "Treaty on Principles Governing the Activities of States in the Exploration and Use of Outer Space, Including the Moon and Other Celestial Bodies," January 27, 1967. http://www.state.gov/t/ac/trt/5181.htm.

"Draft Treaty for the Prevention of Placement of Weapons in Outer Space," February 12, 2008. http://www.ln.mid.ru/brp_4.nsf/e78a48070f128a7b43256999005bcbb3/0d6e0c64d34f8cfac32573ee002d082a?OpenDocument.

European Space Agency. "ConeXpress Factsheet," March 31, 2008. http://telecom.esa.int/telecom/media/document/ORL.pdf.

Garamone, Jim. American Forces Press Service. "CENTCOM Charts Operation Iraqi Freedom Progress," March 25, 2003. http://www.defenselink.mil/news/newsarticle.aspx?id=29230.

General Accounting Office. *MILITARY SPACE OPERATIONS: Planning, Funding, and Acquisition Challenges Facing Efforts to Strengthen Space Control*, September 2002. http://www.gao.gov/new.items/d02738.pdf.

Kan, Shirley. *China's Anti-Satellite Weapon Test*. Congressional Research Service Report for Congress. April 23, 2007. http://www.dtic.mil/cgi-bin/GetTRDoc?AD=ADA468025&Location=U2&doc=GetTRDoc.pdf.

Lawrence Livermore National Laboratory. "HyperSoar." http://www.llnl.gov/str/Carter.html.

National Aeronautics and Space Administration. "Balloon Program Office." http://www.wff.nasa.gov/~code820/missions/missions.html.

National Aeronautics and Space Administration. "Candlestick Rocket Ship," January 23, 2003. http://science.nasa.gov/headlines/y2003/28jan_envirorocket.htm.

National Aeronautics and Space Administration. "Earth's Atmosphere," December 1, 1995. http://liftoff.msfc.nasa.gov/academy/space/atmosphere.html.

National Aeronautics and Space Administration. "Orbital Debris Important Reference Documents," August 20, 2008. http://orbitaldebris.jsc.nasa.gov/library/references.html.

National Aeronautics and Space Administration. "X-37 Demonstrator to Test Future Launch Technologies in Orbit and Reentry Environments," May 2003. http://www.nasa.gov/centers/marshall/pdf/171001main_x37-facts.pdf.

Office of Science and Technology Policy. *U.S. National Space Policy*, August 31, 2006. http://www.ostp.gov/galleries/default-file/Unclassified%20National%20Space%20Policy%20—%20FINAL.pdf.

Rupp, Sheila. "Operationally Responsive Space." *Air Force Print News* May 22, 2007. http://www.kirtland.af.mil/news/story.asp?id=123054292.

United Nations. "GA/SPD/192: RUSSIAN FEDERATION CAUTIONS AGAINST MILITARY DEPLOYMENT IN OUTER SPACE; REITERATES PROPOSAL FOR CONFERENCE TO PREVENT MILITARIZATION," October 17, 2000. http://www.un.org.

United States Air Force. *Air Force Doctrine Document 1*, November 17, 2003. http://www.dtic.mil/doctrine/jel/service_pubs/afdd1.pdf.

United States Air Force. *Air Force Doctrine Document 2–2 Space Operations*, November 27, 2006. http://www.dtic.mil/doctrine/jel/service_pubs/afdd2_2.pdf.

United States Air Force. *Air Force Doctrine Document 2–2.1 Counter-Space*, August 2, 2004. http://www.dtic.mil/doctrine/jel/service_pubs/afdd2_2_1.pdf.

United States Air Force. "Fact Sheet: Air Force Research Laboratory Directed Energy Directorate Fact Sheets." http://www.kirtland.af.mil/afrl_de/factsheets/index.asp.

United States Air Force. "Fact Sheet: Air Force Space Command." http://www.af.mil/factsheets/factsheet.asp?fsID=155.

United States Air Force. "Fact Sheet: Defense Meteorological Satellite Program." http://www.peterson.af.mil/library/factsheets/factsheet.asp?id=8403.

United States Air Force. "Fact Sheet: Defense Satellite Communications System." http://www.losangeles.af.mil/library/factsheets/factsheet.asp?id=5322.

United States Air Force. "Fact Sheet: Defense Support Program Satellites." http://www.losangeles.af.mil/library/factsheets/factsheet.asp?id=5323.

United States Air Force. "Fact Sheet: Global Positioning System." http://www.losangeles.af.mil/library/factsheets/factsheet.asp?id=5325.

United States Air Force. "Fact Sheet: Ground-Based Electro-Optical Deep Space Surveillance." http://www.af.mil/factsheets/factsheet.asp?fsID=170&page=1.

United States Air Force. "Factsheet: Joint Direct Attack Munition GBU-31/32/38," November 2007. http://www.af.mil/factsheets/factsheet.asp?id=108.

United States Air Force. "Factsheet: LGM-30 Minuteman III," November 2006. http://www.af.mil/factsheets/factsheet.asp?id=113.
United States Air Force. "Fact Sheet: 16th Space Control Squadron." http://www.losangeles.af.mil/library/factsheets/factsheet.asp?id=5321.
United States Air Force. "Fact Sheet: Space-Based Infrared Systems." http://www.losangeles.af.mil/library/factsheets/factsheet.asp?id=5330.
United States Air Force. "Program Elements FY2009—Counterspace Systems," February 2008. http://www.js.pentagon.mil/descriptivesum/Y2009/AirForce/0604421F.pdf.
United States Defense Advanced Research Projects Agency. "Orbital Express." http://www.darpa.mil/orbitalexpress/index.html.
United States Department of Defense. *Joint Publication 3–14: Joint Doctrine for Space Operations,* August 9, 2002. http://www.dtic.mil/doctrine/jel/new_pubs/jp3_14.pdf.
United States, Missile Defence Agency. "Fact Sheets." http://www.mda.mil/mdalink/html/factsheet.html.
United States Strategic Command. *Space Missions,* March 2004. http://www.stratcom.mil/factsheetshtml/spacemissions.htm.
U.S. Congress, Office of Technology Assessment. *Anti-Satellite Weapons, Countermeasures, and Arms Control,* OTA-ISC-281. Washington, DC: U.S. Congress, Office of Technology Assessment, September 1985.
U.S. Congress, Office of Technology Assessment. *Arms Control in Space: Workshop Proceedings,* OTA-BP-ISC-28. Washington, DC: U.S. Congress, Office of Technology Assessment, May 1984.
U.S. Congress, Office of Technology Assessment. *Ballistic Missile Defense Technologies,* OTA-ISC-254. Washington, DC: U.S. Government Printing Office, September 1985.
U.S. Congress, Office of Technology Assessment. *Directed Energy Missile Defense in Space–A Background Paper,* OTA-BP-ISC-26. Washington, DC: U.S. Congress, Office of Technology Assessment, April 1984.
U.S. Space Command. *The Long Range Plan.* Colorado Springs, CO: U.S. Space Command, 1997.

Secondary

AirLaunch LLC. "AirLaunch Fact Sheet/Backgrounder," December 2007. http://www.AirLaunchLLC.com.
Arms Control Association. "U.S. Test-Fires 'MIRACL' at Satellite Reigniting ASAT Weapons Debate," *Arms Control Today,* October 1997. http://www.armscontrol.org/act/1997_10/miracloct.
Axe, David. "Semper Fly: Marines in Space." *Popular Science,* December 2006. http://www.popsci.com/military-aviation-space/article/2006-12/semper-fly-marines-space.
Baucom, Donald R. "The Rise and Fall of Brilliant Pebbles." *The Journal of Social, Political and Economic Studies* 29, no. 2 (Summer 2004): 143–90.
Beason, Doug. *The E-Bomb.* Cambridge, MA: Da Capo Press, 2005.
Belote, Howard D. "The Weaponization of Space." *Airpower Journal* (Spring 2000). http://www.airpower.maxwell.af.mil/airchronicles/apj/apj00/spr00/belote.htm.
Berkowitz, Bruce. *The New Face of War.* Toronto, ON: The Free Press, 2003.
Boeing. *Orbital Express Mission Book.* Seattle, WA: Boeing, 2007. http://www.boeing.com/ids/advanced_systems/orbital/pdf/orbital_express_demosys_00.pdf.

Boeing. "Space Based Space Surveillance (SBSS) System." http://www.boeing.com/defense-space/space/satellite/sbss.html.

Boeing. "XSS Micro-satellite." http://www.boeing.com/defense-space/space/xss/.

Bond, Peter. *Jane's Space Recognition Guide*. London: Harper Collins, 2008.

Boot, Max. "The New American Way of War." *Foreign Affairs* 82, no. 4 (July/August 2003). http://www.foreignaffairs.org/20030701faessay15404/max-boot/the-new-american-way-of-war.html.

Booth, Max. *War Made New: Weapons, Warriors, and the Making of the Modern World*. New York: Gotham, 2007.

British Broadcasting Corporation. "Brazil Vows to Pursue Space Plan," August 23, 2003. http://news.bbc.co.uk/1/hi/world/americas/3176395.stm.

British Broadcasting Corporation. "Mir Space Station 1986–2001." http://news.bbc.co.uk/hi/english/static/in_depth/sci_tech/2001/mir/default.stm.

British Broadcasting Corporation. "SpaceShipOne Rockets to Success," October 4, 2004. http://news.bbc.co.uk/1/hi/sci/tech/3712998.stm.

Butler, Amy. "USAF Eyes Counter-ASAT System in 2011." *Aviation Week*, March 16, 2008. http://www.aviationweek.com/aw/generic/story_generic.jsp?channel=awst&id=news/aw031708p2.xml&headline=USAF%20Eyes%20Counter-ASAT%20System%20in%202011.

Butrica, Andrew J. *Single Stage to Orbit: Politics, Space Technology, and the Quest for Reusable Rocketry*. Baltimore, MD: The Johns Hopkins University Press, 2003.

Canan, James W. "Controlling the Space Arena." *AIAA—Aerospace America Online*, January 2004. http://www.aiaa.org/aerospace/images/articleimages/pdf/cananjanuary04.pdf.

Chipman, Donald D., "AIRPOWER A New Way of Warfare (Sea Control)." *Airpower Journal* (Fall 1997). http://www.airpower.maxwell.af.mil/airchronicles/apj/apj97/fal97/chipman.html.

Chun, Clayton. *Defending Space: US Anti-Satellite Warfare and Space Weaponry*. Oxford: Osprey Publishing, 2006.

Chun, Clayton. "Shooting Down a 'Star' Program 437, the US Nuclear ASAT System and Present-Day Copycat Killers." Cadre Paper No. 6, Maxwell Air Force Base, Alabama, April 2000. http://www.maxwell.af.mil/au/aul/aupress/catalog/CADRE_Papers/CADRE_Out/Chun_P8.htm.

Clancy, Tom, and Chuck Horner. *Every Man a Tiger*. New York: G.P. Putnam's Sons, 1999.

Collins, John M. *Military Space Forces*. Washington, DC: Pergamon-Brasesey's International Defense Publishers, 1989.

Covault, Craig. "Chinese Test Anti-Satellite Weapon." *Aviation Week & Space Technology*, January 17, 2007. http://www.aviationweek.com/aw/generic/story_generic.jsp?channel=awst&id=news/CHI01177.xml.

Covault, Craig. "USAF To Launch First Spaceplane Demonstrator." *Aviation Week & Space Technology*, August 3, 2008. http://www.aviationweek.com/aw/generic/story.jsp?id=news/aw080408p2.xml&headline=USAF%20To%20Launch%20First%20Spaceplane%20Demonstrator&channel=awst.

Croft, John. "Stealth Satellites—Cold War Myth or Operational Reality?" *C4ISR Journal*, October 4, 2006. http://www.c4isrjournal.com/story.php?F=2034471.

Czysz, Paul, and Claudio Bruno. *Future Spacecraft Propulsion Systems: Enabling Technologies for Space Exploration*. Berlin: Springer, 2006.

Dalbello, Richard. "OpEd: Putting the "Operational" in Operationally Responsive Space." *Space News*, April 10, 2006. http://www.space.com/spacenews/archive06/DalBelloOpEd_041007.html.

Bibliography

DeBlois, Bruce M. "Space Sanctuary: A Viable National Strategy." *Aerospace Power Journal* (Winter 1998). http://www.airpower.maxwell.af.mil/airchronicles/apj/apj98/win98/deblois.html.

DeBlois, Bruce M., Richard L. Garwin, R. Scott Kemp, and Jeremy C. Marwell. "Space Weapons: Crossing the U.S. Rubicon." *International Security* (Fall 2004). http://mitpress.mit.edu/catalog/item/default.asp?ttype=4&tid=26.

Dinerman, Taylor. "Low-Cost Access to Orbit: Space Marines to the Rescue." *The Space Review*, January 9, 2006. http://www.thespacereview.com/article/530/1.

Dolman, Everett C. *Astropolitik*. Portland, OR: Frank Cass, 2002.

Dupont, Daniel G. "Nuclear Explosions in Orbit." *Scientific American* 290, no. 6 (July 2004): 100–107.

Eisenhower Institute. "A European Perspective on Current Trends in Military and Civilian Space," 2004. http://www.eisenhowerinstitute.org/programs/globalpartnerships/fos/newfrontier/parismeeting.htm.

Eisenhower Institute. "A European Perspective on Current Trends in Military and Commercial Space," July 15, 2002. http://www.eisenhowerinstitute.org/events/past_events/old_events/071502MtngRpt.dot.

Emdee, Jeff. "Launch Vehicle Propulsion." *Crosslink* 5, no. 1 (Winter 2004). http://www.aero.org/publications/crosslink/winter2004/03.html.

Fédération Aéronautique Internationale. "100 km Boundary for Astronautics," June 25, 2004. http://www.fai.org/book/view/22.

Federation of American Scientists. "FAS Calls for Alternatives to Weapons in Space," October 8, 2004. http://fas.org/nuke/control/os/.

Fergusson, James, and Stephen James. *Report on Canada, National Security and Outer Space*. Calgary, Alberta: Canadian Defence and Foreign Affairs Institute, 2007.

Folger, Tim. "Shield of Dreams." *Discover* 22, no. 11 (November 2001): 58–67.

Fortescue, Peter, John Stark, and Graham Swinerd, ed. *Spacecraft Systems Engineering*, 3rd ed. Chichester, UK: John Wiley & Sons, 2003.

France, Martin E. B. "Back to the Future: Space Power Theory and A. T. Mahan." *Air & Space Power Chronicles*, August 2000. http://www.airpower.maxwell.af.mil/airchronicles/cc/france1.html.

Freedman. *The Evolution of Nuclear Strategy*, 2nd ed. London: MacMillan, 1989.

Friedman, George, and Meredith Friedman. *The Future of War: Power, Technology and American World Dominance in the Twenty-first Century*. New York: St. Martin's Griffin, 1998.

Friedman, Thomas L. *The World Is Flat*. New York: Farrar, Straus and Giroux, 2005.

Global Security. "Historical Nuclear Weapons." http://www.globalsecurity.org/wmd/systems/nuke-list.htm.

Global Security. "Navstar Global Positioning System." http://www.globalsecurity.org/space/systems/gps.htm.

Global Security. "Nodong-1." http://www.globalsecurity.org/wmd/world/dprk/nd-1.htm.

Global Security. "Project Babylon Supergun / PC-2." http://www.globalsecurity.org/wmd/world/iraq/supergun.htm.

Global Security. "R-36-O / SL-X-? FOBS." http://www.globalsecurity.org/wmd/world/russia/r-36o.htm.

Global Security. "SENIOR CROWN SR-71." http://www.globalsecurity.org/intell/systems/sr-71.htm/.

Godwin, Robert, ed. *Dyna-Soar: Hypersonic Strategic Weapons System*. Burlington, ON: Apogee, 2003.

Harvey, Brian. *China's Space Program: From Conception to Manned Spaceflight*. Chichester, UK: Springer, 2004.
Hays, Peter L., James M. Smith, Alan R. Van Tassel, and Guy M. Walsh, ed. *Spacepower for a New Millennium*. New York: McGraw-Hill, 2000.
Hendrickx, Bart, and Bert Vis. *Energiya-Buran: The Soviet Space Shuttle*. Chichester, UK: Springer, 2007.
Herken, Gregory. *Counsels of War.* New York: Oxford University Press, 1987.
Hilburn, Matt. "Marines Eye Space-Enabled Options for Expeditionary Warfare," *Seapower Magazine* 49, no. 2 (February 2006). http://www.navyleague.org/sea_power/feb06-28.php.
Hitchens, Theresa. "Europe's USAF Counterspace Operation Doctrine: Questions Answered, Questioned Raised." *Center for Defense Information,* October 4, 2004. http://www.cdi.org/program/document.cfm?DocumentID=2504&from_page=../index.cfm.
Hobbs, David. *Space Warfare*. New York: Prentice Hall, 1986.
Jardin, Xeni. "Bird? Plane? UFO? No, Stratellite." *Wired News,* December 23, 2002. http://www.wired.com/news/wireless/0,1382,56961,00.html.
Johnson-Freese, Joan. *Space as a Strategic Asset*. New York: Columbia University Press, 2007.
Klerkx, Greg. *Lost in Space: The Fall of NASA and the Dream of a New Space Age*. New York: Pantheon Books, 2004.
Krepon, Michael, Jeffery Lewis, and Theresa Hitchens. "Weapons in Space." *Arms Control Today,* November 2004. http://www.armscontrol.org/act/2004_11/Krepon.asp.
Lambeth, Benjamin S. *Mastering the Ultimate High Ground: Next Steps in the Military Uses of Space*. Santa Monica, CA: RAND, 2003. http://www.rand.org/publications/MR/MR1649/.
Lerner, Preston. "A few Dreamers Building Rockets in Work-Shops." *Popular Science* 262, no. 5 (May 2003): 56–64.
Lewis, Jeffrey. "Autonomous Proximity Operations: A Coming Collision in Orbit?" March 11, 2004. http://www.cissm.umd.edu/papers/files/autonomous_proximity.pdf.
Lewis, Jeffrey. "Space Weapons in US Defense Planning." *INESAP Information Bulletin,* April 2004. http://www.inesap.org/bulletin23/art03.htm.
Lockwood, Jonathan S. "Space Control Versus Space Denial in 21st Century Warfare: Achilles Heel of the RMA (Revolution in Military Affairs)?" *Defense & Foreign Affairs Strategic Policy* 28, no. 8 (2000): 4.
Logsdon, John M. "Just Say Wait to Space Power." *Issues in Science and Technology* 17, no. 3 (Spring 2001). http://www.issues.org/17.3/p_logsdon.htm.
Lupton, David E. *On Space Warfare*. Maxwell Air Force Base, AL: Air University Press, 1998. http://www.airpower.maxwell.af.mil/airchronicles/apj/apj98/win98/deblois.html.
McDonough, Thomas R. *Space The Next Twenty-Five Years,* rev. and updated ed. Toronto, ON: John Wiley & Sons, 1989.
McElyea, Tim. *A Vision of Future Space Transportation*. Burlington, Onatario: Apogee Books, 2003.
McKinley, Cynthia A. S. "The Guardians of Space." *Airpower Journal* (Spring 2000). http://www.airpower.maxwell.af.mil/airchronicles/apj/apj00/spr00/mckinley.htm.
McKitrick, Jeffrey, James Blackwell, Fred Littlepage, George Kraus, Richard Blanchfield, and Dale Hill. "The Revolution in Military Affairs." *Battlefield of the Future*. http://www.airpower.maxwell.af.mil/airchronicles/battle/ov-4.html.
Morton, Oliver. "Europe's New Air War." *Wired Magazine,* August 2002. http://wired.com./wired/archive/10.08/airwar.html.

Mowthorpe, Matthew. *The Militarization and Weaponization of Space*. Toronto, ON: Lexington Books, 2004.
Mueller, Karl P. "Totem and Taboo: Depolarizing the Space Weaponization Debate." Paper based on presentation given to Weaponization of Space Project of the Eliot School of International Affairs Space Policy Institute and Security Policy Studies Program, George Washington University, December 3, 2001. http://www.gwu.edu/~spi/space forum/TotemandTabooGWUpaperRevised%5B1%5D.pdf.
Neufeld, Michael J. *Von Braun: Dreamer of Space, Engineer of War*. New York: Vintage Books, 2008.
Oberg, James. "The War of Words over War in Space," April 16, 2004. http://msnbc.msn.com/id/4732874.
O'Hanlon, Michael E. *Neither Star Wars Nor Sanctuary: Constraining the Military Uses of Space*. Washington, DC: Brookings Institution Press, 2004.
Perry, Walter L., and Marc Dean Millot. *Issues from the 1997 Army After Next Winter Wargame*. Santa Monica, CA: RAND Corporation, 1998. http://www.rand.org/pubs/monograph_reports/MR988/.
Petersen, Steven R. *Space Control and the Role of Antisatellite Weapons* (Research Report No.AU-ARI-90-7). Maxwell Air Force Base, AL: Air University Press, May 1991.
Phillips, Terry H. "A Common Aero Vehicle (CAV) Model, Description, and Employment Guide," January 27, 2003. http://www.dtic.mil/dticasd/sbir/sbir041/srch/af031a.doc.
Piscopo, Paul F. *National Aerospace Initiative Update Turning Goals Into Reality*. Washington, DC: National Academy Press, 2004.
Polmar, Norman, and K. J. Moore. *Cold War Submarines*. Washington, DC: Brassey's, 2004.
Rich, Ben R., and Leo Janos. *Skunk Works*. New York: Little, Brown and Company, 1994.
Rife, Shawn P. "On Space-Power Separatism." *Airpower Journal* (Spring 1999). http://www.airpower.maxwell.af.mil/airchronicles/apj/apj99/spr99/rife.html.
Rincon, Paul. British Broadcasting Corporation. "Sat Collision Highlights Growing Threat," February 12, 2009. http://news.bbc.co.uk/2/hi/science/nature/7885750.stm.
Rogers, Lucy. *It's ONLY Rocket Science: An Introduction in Plain English*. New York: Springer, 2008.
Rose, Bill. *Military Space Technology*. Hersham, Surrey, UK: Midland Publishing, 2008.
Ruth, Edward. "That's Why They Call It Rocket Science." *Crosslink* 5, no. 1 (Winter 2004). http://www.aero.org/publications/crosslink/winter2004/02.html.
Scaled Composites. "Tier One Private Manned Space Program." http://www.scaled.com/projects/tierone/faq.htm.
Shachtman, Noah. "Winning-and Losing-the First Wired War." *Popular Science,* June 2006. http://www.popsci.com/scitech/article/2006-06/winning-and-losing-first-wired-war.
Singer, Jeremy. "Satellite Jammer Ready: U.S. Parallel Effort To Thwart Imaging Craft Dropped." *C4ISR Journal* October 19, 2004. http://www.c4isrjournal.com/story.php?F=461040.
Singer, Jeremy. "U.S. Air Force to Upgrade Satcom Jamming System." *Space News,* February 22, 2007. http://www.space.com/spacenews/archive07/countercomm_0219.html.
Sirak, Michael. "US Air Force Eyes 'Near Space' Vehicle." *Jane's Defence Weekly* September 19, 2003. http://www.janes.com/defence/news/jdw/jdw030919_1_n.shtml.
Space Exploration Technologies Corporation—SpaceX. "SpaceX Brochure." http://www.spacex.com/SpaceX_Brochure_V7_All.pdf.
Stares, Paul B. *The Militarization of Space: US Policy, 1945–1984*. Ithaca, NY: Cornell University Press, 1985.

Stares, Paul B. *Space Weapons & US Strategy: Origins & Development.* London: Croom Helm, 1985.
Stephens, Hampton. "Near-Space." *Air Force Magazine Online* 88, no. 7 (July 2005). http://www.afa.org/magazine/July2005/0705near.asp.
Stover, Dawn. "Battlefield Space." *Popular Science* 267, no. 5 (November 2005): 50–57.
Stover, Dawn. "The New War in Space." *Popular Science* 261, no. 3 (September 2002): 40–47.
Sweetman, Bill. "Space Shuttle: The Next Generation." *Popular Science* 262, no. 5 (May 2003): 76–81.
Tajmar, Martin. *Advance Space Propulsion Systems.* Austria: Springer-Verlag/Wien, 2003.
Thomson, Allen. *A Stealth Satellite Sourcebook.* September 29, 2009. http://www.fas.org/spp/military/program/track/stealth.pdf.
United States Air Force, Air University. *Spacecast 2020 Executive Summary.* February 23, 1998. http://www.au.af.mil/Spacecast/monographs/exec-sum.pdf.
Van Allen, Richard E. "The Future of Responsive Space: A Summary of the Results and Papers from the 5th Responsive Space Conference," May 5, 2007. http://www.responsivespace.com/Conferences/RS5/The%20Future%20of%20Responsive%20Space-ASAT-07-08%20.pdf.
William, Martel. *Technological Arsenal.* Washington, DC: Smithsonian, 2001.
Wong, Wilson W. *Weapons in Space: Strategic and Policy Implications. Silver Dart Canadian Aerospace Studies,* vol. 3. Winnipeg, MB: Centre for Defence and Security Studies, University of Manitoba, 2006.
X Prize Foundation. "What is the X PRIZE™?" http://www.xprize.org/press/what.html.

Recommended Web Sites

Air War College Gateway, http://www.au.af.mil/au/awc/awcgate/awcgate.htm
Encyclopedia Astronautica, http://www.astronautix.com
Global Security—Space, http://www.globalsecurity.org/space/index.html
Joint Electronic Library, http://www.dtic.mil/doctrine/index.html
RAND Corporation, http://www.rand.org
Responsive Space, http://www.responsivespace.com
United States Strategic Command, http://www.stratcom.mil

Index

ABM-1 "Galosh." *See* Antiballistic missile systems
Active denial system (ADS), 102
Advance Extremely High Frequency (AEHF) satellite program. *See* Communication satellites
AFP-731 satellite, 78–79
Airborne laser (ABL), 26, 105–106. *See also* Antiballistic missile systems
Air-breathing propulsion, 15, 30–31, 83
Air Launched Miniature Homing Vehicle (ALMHV). *See* Anti-satellite weapon
Almaz (space station), 11, 76, 100
al-Qaeda, 50, 63
Ansari X-Prize, 33. *See also* Spaceship One
Antiballistic missile systems: ABM-1 "Galosh," 9, 11, 48, 94; boost-phase, 99; Brilliant Pebbles, 34, 99, 108; Global Protection Against Limited Strikes, 9, 34; Ground-based Midcourse Defense (GMD), 9, 93, 99; National Missile Defense, 9; Nike, 6, 11, 94, 114; Patriot, 9, 46; Safeguard, 9, 48; Space-Based Kinetic-Kill Vehicle (SBKKV), 99; Standard SM-3, 93, 99. *See also* Airborne laser (ABL); Mid-Infrared Advanced Chemical Laser (MIRACL); Space Based Laser (SBL); Tactical High-Energy Laser (THEL)
Anti-Ballistic Missile Treaty of 1972, 10

Anti-satellite systems: Air Launched Miniature Homing Vehicle (ALMHV), 9, 11, 98–99; Co-Orbital ASAT, 4, 11, 91, 94, 100; Project 437, 94, 114; SC-19, 4–5, 25, 42, 92–94, 98, 100, 108. *See also* Counter Communication System (CCS)
Apollo Program, 3–4, 21, 38, 104, 112
Article IV, Outer Space Treaty. *See* Outer Space Treaty, 1967
Atlas missile. *See* Ballistic missiles
Atmospheric Layers, 14
Automated teller machine (ATM), 56
Automated transfer vehicle (ATV), 91
Autonomous Space Transport Robotic Operations (ASTRO). *See* Orbital Express program

Ballistic missiles: Atlas, 6, 11; Minuteman, 16, 111; Polaris, 6, 10; R-36, 110; R-7, 44; Titan, 6; Trident D-5 42, 111; V-2, 2, 5, 11, 108. *See also* Fractional Orbital Bombardment System (FOBS)
Ballistic Missile Early Warning (BMEW), 67
Barnett, Thomas P. M., 60
B-52, 58, 112
Blue Force Tracking (BFT), 56
Bogotá Declaration of 1976, 8
Bono, Philip, 112
von Braun, Wernher, 6
Brilliant Pebbles. *See* Antiballistic missile systems

Index

Canadarm. *See* Shuttle Remote Manipulator System
Cape Canaveral Air Force Station, 79
CGM-16 Atlas missile. *See* Ballistic missiles
China Aerospace Science and Industry Corporation KT-1 launch vehicle, 92
Clinton, William Jefferson (president), 8, 9, 35, 56, 106
Cobra Dane radar, 67. *See also* Ballistic Missile Early Warning (BMEW)
Cold plasma technology, 107
Collision between Iridium 33 and Kosmos 2251, 70
Collisions in orbit, 70
Committee on the Peaceful Use of Outer Space (COPUOS), United Nations, 6
Common Aero Vehicle (CAV). *See* X-planes
Communication satellites: Advance Extremely High Frequency (AEHF), 61; Defense Satellite Communications Systems (DSCS), 10, 61; Globestar (proposed), 34; Initial Defense Satellite Communications System (IDSCS), 10; Iridium, 21, 34, 42, 59, 61, 70, 82; MILSATCOM, 3, 61; MILSTAR, 10, 61; Molniya, 11, 23, 42; Skynet, 42, 87
ConeXpress Orbital Life Extension Vehicle, 75, 91
Conference of Disarmament's Committee on the Prevention of an Arms Race in Outer Space (PAROS), 6–8
Co-Orbital ASAT. *See* Anti-satellite systems
CORONA satellite program. *See* Reconnaissance satellites
Counter Communication System (CCS), 42, 85, 87

Defense Advanced Research Projects Agency (DARPA), 46, 74, 82, 91, 111
Defense Satellite Communications Systems (DSCS) satellite Program. *See* Communication satellites
Defense Support Program (DSP) satellite program. *See* Early Warning
Digital Spread Spectrum (DSS), 86
Dual-use, 3, 23, 42, 46, 53, 88, 90
Dynasoar, X-20. *See* X-planes

Early Warning: Defense Support Program (DSP), 10, 11, 23, 42, 46, 69; Missile Defense Alarm System (MIDAS), 6, 10; Oko (Eye), 10. *See also* Ballistic Missile Early Warning (BMEW)
Eisenhower, Dwight D. (president), 1, 6, 7
Electrodynamic tethers, 28, 73
Electromagnetic railgun, 99–100
EP-3. *See* Hainan Island incident, 2001
eXperimental Satellite System-11 (XSS-11), 91

Falcon I Launch Vehicle. *See* SpaceX
Force Application and Launch from Continental United States (FALCON), DARPA program, 82–83, 111; Hypersonic Cruise Vehicle (HCV), 111–112; Small Launch Vehicle (SLV), 83
Fractional Orbital Bombardment System (FOBS), 7, 94, 96, 110
Friedman, Thomas L., 60
FY-1C satellite, 92, 93

Gagarin, Yuri, 3
Galileo GPS Satellite Constellation. *See* Global Positioning Systems
Geodesy measurement, 52
Global Navigation Satellite System (GLONASS) Satellite Constellation. *See* Global Positioning Systems
Global Positioning System, 3, 22, 42, 58, 95, 113
Global Positioning Systems: Galileo, 22, 58, 113; Global Navigation Satellite System (GLONASS), 10, 22, 42, 58, 113; Jamming, 85–87; NAVSTAR, 3, 23, 45; Tsyklon, 10
Global Protection Against Limited Strikes (GPALS). *See* Antiballistic missile systems
Global War on Terror (GWOT), 58, 62–63
Globestar (proposed) satellite constellation. *See* Communication satellites
Goddard, Robert, 2
GoogleMaps, 51

Index

Ground-Based Electro-Optical Deep Space Surveillance (GEODSS), 67, 70. *See also* Space Surveillance Network (SSN)
Ground-based midcourse defense (GMD). *See* Antiballistic missile systems
Guided Weapons: cruise missiles, 49, 52, 55, 57, 109, 111–12; Hs-293, 44; Joint Direct Attack Munitions (JDAM), 42, 51–52, 57, 58, 85–87; Small Diameter Bomb (SDB), 57. *See also* X-41 Common Aero Vehicle (CAV)
Gulf War (1991), 9, 45–46, 49, 50, 55, 58

Hague Code of Conduct on Ballistic Missile Proliferation of 2002, 8
Hainan Island incident, 2001, 50
Hearts and Minds operations, 63
HEXAGON satellite. *See* Reconnaissance satellites
HGM-25 Titan missile. *See* Ballistic missiles
History of On-Orbit Satellite Fragmentations, 68
Horizontal Take-Off and Landing (HOTOL), 31
H-II Transfer Vehicle (HTV), 91
Hubble Space Telescope (HST), 26, 54, 70, 74, 77
Hyper-X, X-43. *See* X-planes

India, 4, 11, 43
Initial Defense Satellite Communications System (IDSCS) satellite program. *See* Communication satellites
International Space Station (ISS), 14, 68, 91
Iran, 4, 49, 70
Iridium 33. *See* Collision between Iridium 33 and Kosmos 2251
Iridium satellite phone service. *See* Communication satellites

Jamming: communications. *See* Counter Communication System (CCS); Global Positioning System

von Kármán, Theodore, and the Kármán line, 16

KENNAN/CRYSTAL satellite program. *See* Reconnaissance satellites
Kennedy, John F. (president), 1
Kepler, Johannes, 7, 21
Kessler Syndrome, 25, 69, 100
KH-11 KENNAN/CRYSTAL. *See* Reconnaissance satellites
KH-4 CORONA. *See* Reconnaissance satellites
KH-9 HEXAGON. *See* Reconnaissance satellite
KH-1 CORONA. *See* Reconnaissance satellites
Korolev, Sergei, 44
Kosmos 2251. *See* Collision between Iridium 33 and Kosmos 2251
KT-1. *See* China Aerospace Science and Industry Corporation KT-1 launch vehicle

Laser, 5, 9, 26, 32, 36–37, 101, 104–107; Chemical, 36, 104–105; Communication, 86; Dazzler, 106; Free-electron Laser (FEL), 105; Gamma-ray (GRASER), 104; Guidance, 51–52, 57, 104; Microwave (MASER), 104; propulsion, 106; X-Ray, nuclear bomb pumped, 37, 105. *See also* Airborne Laser (ABL); Mid-Infrared Advanced Chemical Laser (MIRACL); Tactical High-Energy Laser (THEL)
Launch vehicles, 6, 7, 16–17, 21, 26–35, 72, 82–83, 98, 112. *See also* Air-breathing propulsion; Ballistic missiles; Dual-use; Force Application and Launch from Continental United States (FALCON); N1; R-7; Rocket; Safir-2 launch vehicle; Sänger Silbervogel (Silverbird, Amerika Bomber); Saturn V; Space Transportation System (STS); SpaceX; Tsiolkovsky, Konstantin
LGM-30 Minuteman missile. *See* Ballistic missiles
Lockheed Advanced Development Programs (ADP) Division, "Skunk Works," 15, 78
Lockheed Blackbird, aircraft, 15, 24

MapQuest, 51
Micro Unmanned Aerial Vehicle. *See* Unmanned Aerial Vehicle
Midcourse Space Experiment (MSX) satellite. *See also* Space-Based Visible (SBV) sensor
Mid-Infrared Advanced Chemical Laser (MIRACL), 5
MILSATCOM program. *See* Communication satellites
Minuteman missile. *See* Ballistic missiles
Mir Space Station, 36, 91
Missile Defense Alarm System (MIDAS) satellite program. *See* Early Warning
Missile defense, 5–6, 9–10, 26, 34–35, 43, 46, 48, 55, 69, 70, 92–94, 99, 113. *See also* Antiballistic missile systems
MISTY. *See* AFP-731 satellite
Molniya satellite. *See* Communication satellites
Moon Treaty of 1979, 8
Mutual Assured Destruction (MAD), 5, 43, 95
Myrabo, Leik, 32

National Aeronautical and Space Administration (NASA), 1, 3, 16, 46, 68, 69, 73, 74
National Missile Defense. *See* Antiballistic missile system
National Reconnaissance Office (NRO), 6, 53
National Space Policy, United States, 1
NAVSTAR, GPS Constellation. *See* Global Positioning System
Newton, Isaac, 2, 7, 21, 28, 73
Next Generation Satellite/Commodity Spacecraft (NextSat/CSC). *See* Orbital Express program
Nike missile. *See* Antiballistic missile systems
9/11. *See* Terrorist attacks of September 11, 2001
N1, 30
North American Aerospace Defense Command (NORAD), 3

North Korea, 4, 114
Nuclear deterrence, 5, 35, 43–44, 47, 52, 95, 97

Oberth, Herman, 2
Oko (Eye) satellite program. *See* Early Warning
Omid satellite, 70
Operation Desert Storm. *See* Gulf War, 1991
Orbital Express program, 74–75, 91
Orion, Project, 32, 80
Outer Space Treaty of 1967, 4, 7–8, 10, 16, 37–38, 77, 92, 109, 110

Palapa B1 satellite, 85
Patriot missile. *See* Antiballistic missile systems
Polaris missile. *See* Ballistic missiles
Powered hovering, 17
Precision Guided Munitions (PGM). *See* Guided weapons
Project 437. *See* Anti-Satellite Weapon
Prompt Global Strike, 110–12
Pulse detonation engine (PDE), 31–32

Q-Ships, 76

Rapid Attack Identification Detection Reporting System (RAIDRS), 70–71, 87
Reagan, Ronald (president), 1, 9, 83, 99, 105
Reconnaissance satellites: commercial earth imaging, 42, 50, 53; KH-11 KENNAN/CRYSTAL, 42, 54, 79, 87; KH-4 (later-CORONA), 53; KH-1 (CORONA, Discovery), 6, 11, 53, 76; KH-9 (HEXAGON, Big Bird), 53
Refueling, 74–75, 83, 91
Rescue of Astronauts Agreement of 1968, 8
Revolution in Military Affairs (RMA), 41, 43, 47, 62, 67, 101
RIM-161 Standard Missile-3 (SM-3). *See* Antiballistic missile systems
Rocket: Chemical, 28–29, 31–32, 71–72, 98; Hydrazine, 93; Hydrocarbon fuelled, 32; Ion, 72–73; Solid, 28, 33, 73, 92, 106; Thermal, 31, 72; Variable

Specific Impulse Magnetoplasma Rocket (VASIMR), 73
Rocket Equation, Tsiolkovsky's, 29
RQ-1 Predator. See Unmanned Aerial Vehicle
RQ-4A Global Hawk. See Unmanned Aerial Vehicle
R-7 missile (SS-6 SAPWOOD). See Ballistic missiles
R-36 missile (SS-9 SCARP). See Ballistic missiles
Rumsfeld, Donald, 62, 71

Safeguard Program. See Antiballistic missile systems
Safir-2 launch vehicle, 70
SAINT Program, 6
Salyut space station, 76
Sänger Silbervogel (Silverbird, Amerika Bomber), 20
Sänger, Eugene, 20
Satellite and Missile Observation System (SAMOS), 6
Satellite Registration Convention of 1969, 8
Saturn V, 30, 33
SBSS Pathfinder satellite, 70
SC-19. See Anti-Satellite Weapons
Scaled Composites, 33
Second World War, 2, 5, 20, 34, 41, 42, 44, 48, 76, 77
Shuttle Remote Manipulator System (SRMS), 75, 91
Single Stage To Orbit (SSTO), 29, 112
Skylab, 19, 93
Skynet satellite program. See Communication satellites
SM-80 Minuteman missile. See Ballistic missiles
SM-68 Titan missile. See Ballistic missiles
SM-65 Atlas missile. See Ballistic missiles
Solar sails, 28, 74
Southern Cross II, 57. See also Unmanned Aerial Vehicle
Space Based Laser (SBL), 26, 105, 113
Space Based Space Surveillance (SBSS), 70, 77
Space Based Visible, sensor, 67, 70, 77
Space Liability Convention of 1976, 8

Space Race, 4, 11, 38, 44, 108
Space Shuttle. See Space Transportation System
Space Surveillance and Tracking System (STSS), 69
Space Surveillance Network (SSN), 3, 67–70, 79
Space Transportation System (STS), 3, 19, 27, 30, 33–34, 68, 74–76, 80, 83, 91, 110, 111
Space weapons treaties, proposed, 8, 77, 92
Space-Based Infrared-High (SBIRS-H) program, 56, 69
Space-Based Infrared-Low (SBIRS-L) program. See Space Surveillance and Tracking System (STSS)
Space-Based Kinetic-Kill Vehicle (SBKKV). See Antiballistic missile systems
Space-Based Visible (SBV) sensor, 70, 77
Spaceship One, 33
SpaceX, 83
Sputnik One, 3, 7, 17, 36, 108
Standard Missile 3 (SM-3). See Antiballistic missile systems
Starfire Optical Range (SOR), 76
Starfish Prime nuclear test, 94
Stealth, 68, 71, 75–79, 109
Strategic Arms Limitation Treaty of 1972, 76
Strategic Defense Initiative (SDI), 1, 34, 107; Strategic Defense Initiative Office (SDIO), 34
Suborbital space, 14–20, 81, 83, 110–12
Synthetic aperture radar (SAR), 54

Tactical High-Energy Laser (THEL), 105–6
Tactical nuclear weapons, 43, 96–97
Taliban, 50, 58, 63
Terrain Contour Matching (TERCOM), 49–50
Terrorist attacks of September 11, 2001, 62
Three-block war, 62–63
Titan missile. See Ballistic missiles
Townes, Charles, 104

Tourism in space, 44, 81. *See also* Spaceship One
Transit, satellite, 3, 6, 10
Trident missile. *See* Ballistic missiles
Tsiolkovsky, Konstantin, 2, 29
Tsyklon GPS Satellite. *See* Global Positioning System

Ufimtsev, Pyotr Ya, 78
UGM-93 Trident missile. *See* Ballistic missiles
UGM-27 Polaris missile. *See* Ballistic missiles
Unmanned Aerial Vehicle (UAV), 55, 57, 59–60; Global Hawk, 57, 59–60; Micro, 55; Predator, 55, 59–60
Upravlyaemyj Sputnik–Aktivnyj (US-A), 11, 36
USA-193 satellite, 93
U.S. Army (After Next) 1997 Winter War Game, 94
US 51. *See* AFP-731 satellite

US-Iraq War, 2003, 42, 62, 85–86, 95
USS *Lake Erie*, 93
U.S. Space Command Long Range Plan, 47, 66

Van Allen Belts, 22, 25–26, 94
Vandenberg Air Force Base, 79
Very Large Array (VLA) radio telescope, 82
Warfare: atomic age, 43–44; industrial age, 41–42; information age, 41, 43, 45–49, 59–64, 69, 86, 88, 101
Whipple bumper, 80

X-Planes: X-41 Common Aero Vehicle (CAV), 83, 111; X-30, 31, 83, 112; X-37, 83, 111; X-20 Dynasoar, 83. *See also* Force Application and Launch from Continental United States (FALCON); X-43 Hyper-X, 83, 111–12
X-Prize. *See* Ansari X-Prize
Xichang Space Center, 92

About the Authors

WILSON W. S. WONG is a Research Fellow with the Centre for Defence and Securities Studies, Winnipeg, Canada. He has a Master of Arts in Political Studies from the University of Manitoba. This is his second book on the military use of the outer space environment. His first book, *Weapons in Space: Strategic and Policy Implications,* was Volume III in the Centre for Defence and Security Studies' Silver Dart Canadian Aerospace Studies series. In addition to research support, Mr. Wong contributed illustrations to the Canadian Defence and Foreign Affairs Institute's 2007 *Report on Canada, National Security and Outer Space* authored by Dr. James Fergusson and Stephen James. Presently, his ambitions include finding further outlets to combine his interest in technology and public policy.

DR. JAMES FERGUSSON is the Director of the Centre for Defence and Security Studies, Winnipeg Canada, a Professor in the Department of Political Studies at the University of Manitoba, and a Senior Research Fellow with the Canadian Defence and Foreign Affairs Institute. He teaches a range of courses in the areas of international relations, strategic studies, and foreign and defense policy, with an emphasis on Canada. He has recently completed a book, *Déjà vu All Over Again; Canada and Ballistic Missile Defence,* forthcoming from University of British Columbia Press as part of the Canada War Museum Series. His recent publications include "Thinking about a 'Known Unknown': The Past, Present and Future Implications of Strategic Defence," *International Journal,* Winter 2008; and *Canada, National Security and Outer Space,* coauthored with Stephen James and published by the Canadian Defence and Foreign Affairs Institute. He is currently completing a major study on the transformation of the Canada-U.S. defense relationship.

In addition to his academic publications, Dr. Fergusson has been commissioned to write several reports for the Department of National Defence, Canada, and the Department of Foreign Affairs, Canada. He lectures to a wide range of military audiences, including the Canadian Forces. Dr. Fergusson has testified on several occasions to the Standing Committee on Foreign Affairs and International Trade and the Standing Committee on National Defence and Veteran's Affairs on a variety of issues. He has also been a member of the Defence Science Advisory Board and the Defence Industrial Advisory Committee.